U.S. Department
of Transportation
**Federal Aviation
Administration**

M000076187

Effective April 1, 1996

**Change 1    4/28/97**
**Change 2    5/21/97**

# Rotorcraft/
# Helicopter
## For Private and
## Commercial Pilots
# Practical Test
# Standards

**Office of Flight Operations**
Washington, DC 20591

**Reprinted by Aviation Supplies & Academics, Inc.**
Newcastle, WA 98059-3153

**Rotorcraft/Helicopter**
**For Private and Commercial Pilots**
**Practical Test Standards**

Aviation Supplies & Academics, Inc.
7005 132nd Place SE
Newcastle, Washington 98059-3153

Printed in the United States of America

05  04  03            9  8  7  6  5  4  3


ISBN 1-156027-370-4
**ASA-8081-HD**

# PRIVATE PILOT
# ROTORCRAFT/HELICOPTER

## Practical Test Standards

## Section 1

1996

**FLIGHT STANDARDS SERVICE**
**Washington, DC 20591**

# NOTE

Material in FAA-S-8081-15 will be effective April 1, 1996. All previous editions of the Private Pilot – Rotorcraft (Helicopter) Practical Test Standards will be obsolete as of this date.

# FOREWORD

The Private Pilot – Rotorcraft (Helicopter) Practical Test Standards (PTS) book has been published by the Federal Aviation Administration (FAA) to establish the standards for private pilot certification practical tests for the rotorcraft category, helicopter class. FAA inspectors and designated pilot examiners shall conduct practical tests in compliance with these standards. Flight instructors and applicants should find these standards helpful during training and when preparing for the practical test.

---

Thomas C. Accardi
Director, Flight Standards Service

# RECORD OF CHANGES

**Change 1: 4/28/97**
**Reason: Wrong terminology. Changed airplane to aircraft.**

- **HELICOPTER**

  **AREA OF OPERATION: PREFLIGHT PREPARATION**

  **TASK G: MINIMUM EQUIPMENT LIST**

**Change 2: 5/21/97**
**Reason: Under the new rule (8/4/97) the NOTE is not applicable.**

- **HELICOPTER**

  **AREA OF OPERATION: NIGHT OPERATIONS**

  **TASK A: PHYSIOLOGICAL ASPECTS OF NIGHT FLYING**

# CONTENTS

FAA-S-8081-15 **Private**

# INTRODUCTION

The Flight Standards Service of the Federal Aviation Administration (FAA) has developed this practical test book as a standard to be used by FAA inspectors and designated pilot examiners when conducting private pilot – rotorcraft (helicopter) practical tests. Flight instructors are expected to use this book when preparing applicants for practical tests. Applicants should be familiar with this book and refer to these standards during their training.

Information considered directive in nature is described in this practical test book in terms such as "shall" and "must" indicating the actions are mandatory. Guidance information is described in terms such as "should" and "may" indicating the actions are desirable or permissive but not mandatory.

The FAA gratefully acknowledges the valuable assistance provided by a nationwide public "Job Task Analysis" team that developed the knowledge, skills, and abilities that appear in this book. We would also like to thank the many individuals and organizations who contributed their time and talent in assisting with the revision of these practical test standards.

This publication may be obtained from FedWorld through the use of a computer modem or purchased from the Superintendent of Documents, U.S. Government Printing Office, Washington, DC 20402.

Comments regarding this publication should be sent to:

U.S. Department of Transportation
Federal Aviation Administration
Flight Standards Service
Operations Support Branch, AFS-630
P.O. Box 25082
Oklahoma City, OK 73125

## PRACTICAL TEST STANDARD CONCEPT

Federal Aviation Regulations (FAR's) specify the areas in which knowledge and skill must be demonstrated by the applicant before the issuance of a private pilot certificate. The FAR's provide the flexibility to permit the FAA to publish practical test standards containing specific TASKS in which pilot competency must be demonstrated. The FAA will revise this book whenever it is determined that changes are needed in the interest of safety. Adherence to the provisions of the regulations and the practical test standards is mandatory for the evaluation of private pilot applicants.

## PRIVATE PILOT ROTORCRAFT PRACTICAL TEST BOOK DESCRIPTION

This test book contains the following private pilot practical test standards:

    **Section 1**    Rotorcraft, Helicopter – Private Pilot

The Private Pilot Rotorcraft Practical Test Standards include the AREAS OF OPERATION and TASKS for the issuance of an initial private pilot certificate and for the addition of category and/or class ratings to that certificate.

## PRACTICAL TEST STANDARD DESCRIPTION

AREAS OF OPERATION are phases of the practical test arranged in a logical sequence within this standard. They begin with preflight preparation and end with post-flight procedures. The examiner, however, may conduct the practical test in any sequence that results in a complete and efficient test.

The REFERENCE identifies the publication(s) that describe(s) the TASK. Descriptions of TASKS are not included in the standards because this information can be found in the reference list. Publications other than those listed may be used, if their content conveys substantially the same meaning as the listed publications.

Reference list:

| | |
|---|---|
| **FAR Part 43** | Maintenance, Preventive Maintenance, Rebuilding, and Alteration |
| **FAR Part 61** | Certification: Pilots and Flight Instructors |
| **FAR Part 67** | Medical Standards and Certification |
| **FAR Part 91** | General Operating and Flight Rules |
| **NTSB Part 830** | Notification and Reporting of Aircraft Accidents and Incidents |
| **AC 00-2** | Advisory Circular Checklist |
| **AC 00-6** | Aviation Weather |
| **AC 00-45** | Aviation Weather Services |
| **AC 61-13** | Basic Helicopter Handbook |
| **AC 61-21** | Flight Training Handbook |
| **AC 61-23** | Pilot's Handbook of Aeronautical Knowledge |
| **AC 61-65** | Certification: Pilots and Flight Instructors |
| **AC 61-84** | Role of Preflight Preparation |
| **AC 90-48** | Pilots' Role in Collision Avoidance |
| **AC 90-87** | Helicopter Dynamic Rollover |
| **AC 91-13** | Cold Weather Operation of Aircraft |
| **AC 91-23** | Pilot's Weight and Balance Handbook |
| **AC 91-32** | Safety In and Around Helicopters |
| **AC 91-42** | Hazards of Rotating Propeller and Helicopter Rotor Blades |
| **AIM** | Aeronautical Information Manual |
| **AFD** | Airport Facility Directory |
| **NOTAM's** | Notices to Airmen, Helicopter Flight Manuals, Industry Related Manuals |

The Objective lists the important elements that must be satisfactorily performed to demonstrate competency in a TASK. The Objective includes:

1. specifically what the applicant should be able to do;
2. the conditions under which the TASK is to be performed; and
3. the acceptable standards of performance.

## USE OF THE PRACTICAL TEST STANDARDS BOOK

The FAA requires that all practical tests be conducted in accordance with the appropriate Private Pilot Practical Test Standards and the policies set forth in this INTRODUCTION. Private pilot applicants shall be evaluated in **ALL** TASKS included in the AREAS OF OPERATION of the appropriate practical test standard.

In preparation for the practical test, the examiner shall develop a written "plan of action." The "plan of action" shall include all TASKS in each AREA OF OPERATION.

The examiner is not required to follow the precise order in which the AREAS OF OPERATION and TASKS appear in this book. The examiner may change the sequence or combine TASKS with similar Objectives to meet the orderly and efficient flow of the practical test. For example, lost procedures may be combined with radio navigation. The examiner's "plan of action" shall include the order and combination of TASKS to be demonstrated by the applicant in a manner that will result in an efficient and valid test.

Examiners shall place special emphasis upon areas of aircraft operation that are most critical to flight safety. Among these areas are precise aircraft control and sound judgment in decision making. Although these areas may or may not be shown under each TASK, they are essential to flight safety and shall receive careful evaluation throughout the practical test. THE EXAMINER SHALL ALSO EMPHASIZE WAKE TURBULENCE AVOIDANCE, LOW LEVEL WIND SHEAR, INFLIGHT COLLISION AVOIDANCE, RUNWAY INCURSION AVOIDANCE, AND CHECKLIST USAGE.

The examiner is expected to use good judgment in the performance of simulated emergency procedures. The use of the safest means for simulation is expected. Consideration must always be given to local conditions (both meteorological and topographical), the examiner's level of performance at the time of the test, as well as the applicant's, ATC workload, and the relative condition of the aircraft used. If the procedure being evaluated would put the maneuver in jeopardy of safe operation, it is expected that the applicant shall simulate that portion of the maneuver, i.e. - engine governor, trim system malfunction, etc., unless otherwise indicated by the NOTE in a particular AREA OF OPERATION or TASK.

## PRIVATE PILOT ROTORCRAFT PRACTICAL TEST PREREQUISITES

An applicant for the private pilot rotorcraft practical test is required by Federal Aviation Regulations to:

1. pass the appropriate private pilot knowledge test since the beginning of the 24th month before the month in which the practical test is taken;
2. obtain the applicable instruction and aeronautical experience prescribed for the private pilot certificate or training sought;
3. hold at least a current third-class medical certificate issued under FAR Part 67;

4. be at least 17 years of age, and;
5. obtain a written statement from an appropriately certificated flight instructor certifying that the applicant has been given flight instruction in preparation for the practical test within 60 days preceding the date of application. The statement shall also state that the instructor finds the applicant competent to pass the practical test and that the applicant has satisfactory knowledge of the subject area(s) in which a deficiency was indicated by the airman knowledge test report.

AC 61-65, Certification: Pilots and Flight Instructors, states that the instructor may sign the instructor's recommendation on the reverse side of FAA Form 8710-1, Airman Certificate and/or Rating Application, in lieu of the previous statement, provided all appropriate FAR Part 61 requirements are substantiated by reliable records.

## AIRCRAFT AND EQUIPMENT REQUIRED FOR THE PRACTICAL TEST

The private pilot applicant is required by FAR Section 61.45 to provide an airworthy, certificated aircraft for use during the practical test. This section further requires that the aircraft:

1. have fully functioning dual controls, except as provided in this FAR Section; and
2. be capable of performing **ALL** appropriate TASKS for the private pilot certificate and have no operating limitations that prohibit the performance of those TASKS.

## METRIC CONVERSION INITIATIVE

To assist the pilots in understanding and using the metric measurement system, the practical test standards refer to the metric equivalent of various altitudes throughout. The inclusion of meters is intended to familiarize pilots with its use. The metric altimeter is arranged in 10 meter increments; therefore, when converting from feet to meters, the exact conversion, being too exact for practical purposes, is rounded to the nearest 10 meter increment or even altitude as necessary.

## POSITIVE EXCHANGE OF FLIGHT CONTROLS

During the practical test, there must always be a clear understanding of who has control of the aircraft. Prior to the flight, a briefing should be conducted that includes the procedure for the exchange of flight controls. A positive three-step process in the exchange of flight controls between pilots is a proven procedure and one that is recommended.

When the examiner wishes to take the controls to allow the applicant to adjust the seat, headset, etc., he/she will say "I have the controls." The applicant will acknowledge immediately by saying, "You have the controls." The examiner again says, "I have the controls." When control is returned to the applicant, follow the same procedure. A visual check is recommended to verify that the exchange has occurred. There should never be any doubt as to who is flying the aircraft.

## USE OF DISTRACTIONS DURING PRACTICAL TESTS

Numerous studies indicate that many accidents have occurred when the pilot has been distracted during critical phases of flight. To evaluate the pilot's ability to utilize proper control technique while dividing attention both inside and/or outside the cockpit, the examiner shall cause a realistic distraction during the **flight** portion of the practical test to evaluate the applicant's ability to divide attention while maintaining safe flight.

## APPLICANT'S USE OF PRESCRIBED CHECKLISTS

Throughout the practical test, the applicant is evaluated on the use of the prescribed checklist. The situation may be such that the use of the checklist while accomplishing the elements of the objective would be either unsafe or impractical, especially in a single-pilot operation. In this case, it may be more prudent to review the checklist after the elements have been met.

## CREW RESOURCE MANAGEMENT (CRM)

CRM "...refers to the effective use of ALL available resources; human resources, hardware, and information." Human resources "...includes all other groups routinely working with the cockpit crew (or pilot) who are involved in decisions that are required to operate a flight safely. These groups include, but are not limited to: dispatchers, cabin crewmembers, maintenance personnel, and air traffic controllers." CRM is not a single TASK, it is a set of knowledge and skill competencies that must be evident in all TASKS in this PTS as applied to either single pilot or a crew operation.

## FLIGHT INSTRUCTOR RESPONSIBILITY

An appropriately rated flight instructor is responsible for training the private pilot applicant to acceptable standards in *all* subject matter areas, procedures, and maneuvers included in the TASKS within the appropriate private pilot practical test standard. Because of the impact of their teaching activities in developing safe, proficient pilots, flight instructors should exhibit a high level of knowledge, skill, and ability.

Additionally, the flight instructor must certify that the applicant is able to perform safely as a private pilot and is competent to pass the required practical test.

Throughout the applicant's training, the flight instructor is responsible for emphasizing the performance of effective visual scanning, collision avoidance, and runway incursion avoidance procedures.

## EXAMINER[1] RESPONSIBILITY

The examiner conducting the practical test is responsible for determining that the applicant meets the acceptable standards of knowledge and skill of each TASK within the appropriate practical test standard. Since there is no formal division between the **oral** and **skill** portions of the practical test, this becomes an ongoing process throughout the test. To avoid unnecessary distractions, oral questioning should be used judiciously at all times, especially during the flight portion of the practical test.

Examiners shall test to the greatest extent practicable the applicant's correlative abilities rather than mere rote enumeration of facts throughout the practical test.

Throughout the flight portion of the practical test, the examiner shall evaluate the applicant's use of visual scanning and collision avoidance procedures.

## SATISFACTORY PERFORMANCE

Satisfactory performance to meet the requirements for certification is based on the applicant's ability to safely:

1. perform the approved areas of operation for the certificate or rating sought within the approved standards;
2. demonstrate mastery of the aircraft with the successful outcome of each task performed never seriously in doubt;
3. demonstrate sound judgment aeronautical decision making and skill competencies in CRM.

---

[1] The word "examiner" denotes either the FAA inspector or FAA designated pilot examiner who conducts the practical test.

## UNSATISFACTORY PERFORMANCE

If, in the judgment of the examiner, the applicant does not meet the standards of performance of any TASK performed, the associated AREA OF OPERATION is failed and therefore, the practical test is failed. The examiner or applicant may discontinue the test any time after the failure of an AREA OF OPERATION makes the applicant ineligible for the certificate or rating sought. The test will be continued ONLY with the consent of the applicant. If the test is either continued or discontinued, the applicant is entitled credit for only those TASKS satisfactorily performed. However, during the retest and at the discretion of the examiner, any TASK may be re-evaluated including those previously passed.

Typical areas of unsatisfactory performance and grounds for disqualification are:

1. Any action or lack of action by the applicant that requires corrective intervention by the examiner to maintain safe flight.
2. Failure to use proper and effective visual scanning techniques to clear the area before and while performing maneuvers.
3. Consistently exceeding tolerances stated in the Objectives.
4. Failure to take prompt corrective action when tolerances are exceeded.

When a disapproval notice is issued, the examiner will record the applicant's unsatisfactory performance and TASKS not completed in terms of AREA OF OPERATIONS appropriate to the practical test conducted.

## USE OF RATING TASKS TABLES

If an applicant already holds a private pilot certificate, use the appropriate table at the beginning of each section, to determine which TASKS are required on the practical test. However, at the discretion of the examiner, the applicant's competence in any TASK may be evaluated, if indications of the applicant's performance suggests that such action is appropriate.

If the applicant holds more than one category or class rating at the private level, and the table indicates differing required TASKS, the "least restrictive" entry applies. For example, if "ALL" and "NONE" are indicated for one AREA OF OPERATION, the "NONE" entry applies. If "B" and "B, C" are indicated, the "B" entry applies.

# CONTENTS: SECTION 1

## Addition of a Rotorcraft/Helicopter rating
## to an existing Private Pilot Certificate

| Area of Opera-tion | Required TASKS are indicated by either the TASK letter(s) that apply(s) or an indication that all or none of the TASKS must be tested. | | | | | | | | |
|---|---|---|---|---|---|---|---|---|---|
| | PRIVATE PILOT RATING(S) HELD | | | | | | | | |
| | ASEL | ASES | AMEL | AMES | RG | Non-Power Glider | Power Glider | Free Balloon | Airship |
| I | E,F,G | E,F,G | E,F,G | E,F,G | E,F,G | E,F,G | E,F,G | E,F,G | E,F,G |
| II | ALL | ALL | ALL | ALL | ALL | ALL | ALL | ALL | ALL |
| III | B,C | B,C | B,C | B,C | ALL | ALL | ALL | ALL | B,C |
| IV | ALL | ALL | ALL | ALL | ALL | ALL | ALL | ALL | ALL |
| V | ALL | ALL | ALL | ALL | ALL | ALL | ALL | ALL | ALL |
| VI | ALL | ALL | ALL | ALL | ALL | ALL | ALL | ALL | ALL |
| VII | NONE | NONE | NONE | NONE | B | B,C,D | B,C,D | B,C,D | NONE |
| VIII | ALL | ALL | ALL | ALL | ALL | ALL | ALL | ALL | ALL |
| IX | NONE | NONE | NONE | NONE | NONE | ALL | ALL | ALL | ALL |
| X | ALL | ALL | ALL | ALL | ALL | ALL | ALL | ALL | ALL |
| | | | | | | | | | |

## TASK VS. SIMULATION DEVICE CREDIT

Examiners conducting the Private Pilot — Helicopter Practical Tests with simulation devices should consult appropriate documentation to ensure that the device has been approved for training. The documentation for each device should reflect that the following activities have occurred:

1. The device must be evaluated, determined to meet the appropriate standards, and assigned the appropriate qualification level by the National Simulator Program Manager. The device must continue to meet qualification standards through continuing evaluations as outlined in the appropriate advisory circular (AC). For helicopter simulators, AC 120-63 (as amended), Helicopter Simulator Qualification, will be used.

2. The FAA must approve the device for specific TASKS.

3. The device must continue to support the level of student or applicant performance required by this PTS.

**NOTE:** Users of the following chart are cautioned that use of the chart alone is incomplete. The description and objective of each task as listed in the body of the PTS, including all notes, must also be incorporated for accurate simulation device use.

## USE OF CHART    (2 OF 4)

**X**    Creditable.

**X1**  Creditable only if accomplished in conjunction with a
         running takeoff or running landing, as appropriate.

**NOTE:**  1. The helicopter may be used for all tasks.
           2. Level C simulators may be used as indicated only if the applicant
              meets established pre-requisite experience requirements.
           3. Level A helicopter simulator standards have not been defined.
           4. Helicopter flight training devices have not been defined.

## II. Preflight Procedures

A. Preflight Inspection (Cockpit Only)
B. Cockpit Management
C. Engine Starting and Rotor Engagement
D. Before Takeoff Check

## III. Airport and Heliport Operations

A. Radio Communications and ATC Light Signals
B. Traffic Patterns
C. Airport and Heliport Markings and Lighting

## IV. Hovering Maneuvers

A. Vertical Takeoff and Landing
B. Slope Operations
C. Surface Taxi
D. Hover Taxi
E. Air Taxi

## V. Takeoffs, Landings, and Go-Arounds

A. Normal and Crosswind Takeoff and Climb
B. Normal and Crosswind Approach
C. Maximum Performance Takeoff and Climb
D. Steep Approach
E. Rolling Takeoff
F. Shallow Approach and Running /Roll-On
G. Go-Around

| FLIGHT TASK<br>Areas of Operation: | 1 | 2 | 3 | 4 | 5 | 6 | 7 | A | B | C | D |
|---|---|---|---|---|---|---|---|---|---|---|---|
| **VI. Performance Maneuvers** | | | | | | | | | | | |
| A. Rapid Deceleration | | | | | | | | | | | |
| B. Straight In Autorotations | | | | | | | | | | | |
| **VII. Navigation** | | | | | | | | | | | |
| A. Pilotage and Dead Reckoning | | | | | | | | | | | |
| B. Radio Navigation and Radar Services | | | | | | | | | | | |
| C. Diversion | | | | | | | | | | | |
| D. Lost Procedures | | | | | | | | | | | |
| **VIII. Emergency Operations** | | | | | | | | | | | |
| A. Power Failure at a Hover | | | | | | | | | | X | X |
| B. Power Failure at Altitude | | | | | | | | | | X | X |
| C. Systems and Equipment Malfunctions | | | | | | | | | | X | X |
| D. Settling-With-Power | | | | | | | | | | X | X |
| E. Low Rotor RPM Recovery | | | | | | | | | | X | X |
| F. Dynamic Rollover | | | | | | | | | | | |
| G. Ground Resonance | | | | | | | | | | | |
| H. Low G Conditions | | | | | | | | | | | |
| I. Emergency Equipment and Survival Gear | | | | | | | | | | | |
| **X. Postflight Procedures** | | | | | | | | | | | |
| A. After Landing and Securing | | | | | | | | | | | |

# APPLICANT'S PRACTICAL TEST CHECKLIST
## (HELICOPTER)
### APPOINTMENT WITH EXAMINER:

**EXAMINER'S NAME**_____

**LOCATION** _____

**DATE/TIME** _____

## ACCEPTABLE AIRCRAFT

- ☐ Aircraft Documents:
    Airworthiness Certificate
    Registration Certificate
    Operating Limitations
- ☐ Aircraft Maintenance Records:
    Logbook Record of Airworthiness Inspections
    and AD Compliance
- ☐ Pilot's Operating Handbook, FAA-Approved
  Helicopter Flight Manual
- ☐ FCC Station License

## PERSONAL EQUIPMENT

- ☐ Current Aeronautical Charts
- ☐ Computer and Plotter
- ☐ Flight Plan Form
- ☐ Flight Logs
- ☐ Current AIM, Airport Facility Directory, and Appropriate
    Publications

## PERSONAL RECORDS

- ☐ Identification - Photo/Signature ID
- ☐ Pilot Certificate
- ☐ Current and Appropriate Medical Certificate
- ☐ Completed FAA Form 8710-1, Airman Certificate and/or
    Rating Application with Instructor's Signature (if applicable)
- ☐ AC Form 8080-2, Airman Written Test Report, or
    Computer Test Report
- ☐ Pilot Logbook with Appropriate Instructor Endorsements
- ☐ FAA Form 8060-5, Notice of Disapproval (if applicable)
- ☐ Approved School Graduation Certificate (if applicable)
- ☐ Examiner's Fee (if applicable)

# EXAMINER'S PRACTICAL TEST CHECKLIST

## (HELICOPTER)

**APPLICANT'S NAME** _____

**LOCATION** _____

**DATE/TIME** _____

### I. PREFLIGHT PREPARATION

- ☐ **A.** CERTIFICATES AND DOCUMENTS
- ☐ **B.** WEATHER INFORMATION
- ☐ **C.** CROSS-COUNTRY FLIGHT PLANNING
- ☐ **D.** NATIONAL AIRSPACE SYSTEM
- ☐ **E.** PERFORMANCE AND LIMITATIONS
- ☐ **F.** OPERATION OF SYSTEMS
- ☐ **G.** MINIMUM EQUIPMENT LIST
- ☐ **H.** AEROMEDICAL FACTORS

### II. PREFLIGHT PROCEDURES

- ☐ **A.** PREFLIGHT INSPECTION
- ☐ **B.** COCKPIT MANAGEMENT
- ☐ **C.** ENGINE STARTING AND ROTOR ENGAGEMENT
- ☐ **D.** BEFORE TAKEOFF CHECK

### III. AIRPORT AND HELIPORT OPERATIONS

- ☐ **A.** RADIO COMMUNICATIONS AND ATC LIGHT SIGNALS
- ☐ **B.** TRAFFIC PATTERNS
- ☐ **C.** AIRPORT AND HELIPORT MARKINGS AND LIGHTING

### IV. HOVERING MANEUVERS

- ☐ **A.** VERTICAL TAKEOFF AND LANDING
- ☐ **B.** SLOPE OPERATIONS
- ☐ **C.** SURFACE TAXI
- ☐ **D.** HOVER TAXI
- ☐ **E.** AIR TAXI

## V. TAKEOFFS, LANDINGS, AND GO-AROUNDS

- ☐ **A.** NORMAL AND CROSSWIND TAKEOFF AND CLIMB
- ☐ **B.** NORMAL AND CROSSWIND APPROACH
- ☐ **C.** MAXIMUM PERFORMANCE TAKEOFF AND CLIMB
- ☐ **D.** STEEP APPROACH
- ☐ **E.** ROLLING TAKEOFF
- ☐ **F.** SHALLOW APPROACH AND RUNNING/ROLL-ON LANDING
- ☐ **G.** GO-AROUND

## VI. PERFORMANCE MANEUVERS

- ☐ **A.** RAPID DECELERATION
- ☐ **B.** STRAIGHT IN AUTOROTATION

## VII. NAVIGATION

- ☐ **A.** PILOTAGE AND DEAD RECKONING
- ☐ **B.** RADIO NAVIGATION AND RADAR SERVICES
- ☐ **C.** DIVERSION
- ☐ **D.** LOST PROCEDURES

## VIII. EMERGENCY OPERATIONS

- ☐ **A.** POWER FAILURE AT A HOVER
- ☐ **B.** POWER FAILURE AT ALTITUDE
- ☐ **C.** SYSTEMS AND EQUIPMENT MALFUNCTIONS
- ☐ **D.** SETTLING-WITH-POWER
- ☐ **E.** LOW ROTOR RPM RECOVERY
- ☐ **F.** DYNAMIC ROLLOVER
- ☐ **G.** GROUND RESONANCE
- ☐ **H.** LOW G CONDITIONS
- ☐ **I.** EMERGENCY EQUIPMENT AND SURVIVAL GEAR

## IX. NIGHT OPERATIONS

- ☐ **A.** PHYSIOLOGICAL ASPECTS OF NIGHT FLYING
- ☐ **B.** LIGHTING AND EQUIPMENT FOR NIGHT FLYING

## X. POST-FLIGHT PROCEDURES

- ☐ AFTER LANDING AND SECURING

# I. AREA OF OPERATION: PREFLIGHT PREPARATION

## A. TASK: CERTIFICATES AND DOCUMENTS

REFERENCES: FAR Parts 43, 61, 67, 91; AC 61-13, AC 61-23; Helicopter Flight Manual.

**Objective.** To determine that the applicant:

1. Exhibits knowledge of the elements related to certificates and documents by explaining—

   a. pilot certificate privileges and limitations.
   b. medical certificate class and duration.
   c. pilot logbook or flight records.

2. Exhibits knowledge of the elements related to certificates and documents by locating and explaining—

   a. airworthiness and registration certificates.
   b. operating limitations, placards, and instrument markings.
   c. weight and balance data and equipment list.
   d. airworthiness directives, compliance records, maintenance requirements, and appropriate records.

## B. TASK: WEATHER INFORMATION

REFERENCES: AC 00-6, AC 00-45, AC 61-23, AC 61-84; AIM.

**Objective.** To determine that the applicant:

1. Exhibits knowledge of the elements related to weather information from various sources with emphasis on—

   a. PIREP's.
   b. SIGMET's and AIRMET's.
   c. wind shear reports.
   d. use of weather reports and forecasts.

2. Makes a competent "go/no-go" decision based on available weather information.

## C. TASK: CROSS-COUNTRY FLIGHT PLANNING

REFERENCES: AC 61-21, AC 61-23, AC 61-84; Navigation Charts; Airport/Facility Directory; NOTAM's; AIM.

**Objective.** To determine that the applicant:

1. Exhibits knowledge of the elements related to cross-country flight planning by presenting and explaining a pre-planned VFR cross-country flight, as previously assigned by the examiner. The flight plan shall be to the first fuel stop necessary, based on maximum allowable passenger, baggage, and/or cargo loads using real-time weather.
2. Uses appropriate and current aeronautical charts.
3. Properly identifies airspace, obstacles, and terrain features, including discussion of wire strike avoidance techniques.
4. Selects easily identifiable en route checkpoints.
5. Selects the most favorable altitudes, considering weather conditions and equipment capabilities.
6. Computes headings, flight time, and fuel requirements.
7. Selects appropriate navigation systems/facilities and communication frequencies.
8. Extracts and applies pertinent information from NOTAM's, Airport/Facility Directory, and other flight publications.
9. Completes a navigation log and simulates filing a VFR flight plan.

## D. TASK: NATIONAL AIRSPACE SYSTEM

REFERENCES: FAR Part 91; AIM.

**Objective.** To determine that the applicant exhibits knowledge of the elements related to the National Airspace System by explaining:

1. Basic VFR Weather Minimums – for all classes of airspace.
2. Airspace classes – their boundaries, pilot certification, and helicopter equipment requirements for the following—

   a. Class A.
   b. Class B.
   c. Class C.
   d. Class D.
   e. Class E.
   f. Class G.

3. Special use airspace and other airspace areas.

## E. TASK:   PERFORMANCE AND LIMITATIONS

REFERENCES:   AC 61-13, AC 61-84, AC 91-23; Helicopter Flight Manual.

**Objective.** To determine that the applicant:

1. Exhibits knowledge of the elements related to performance and limitations by explaining the use of charts, tables, and data to determine performance and the adverse effects of exceeding limitations.
2. Computes weight and balance, including adding, removing, and shifting weight.
3. Determines if the weight and center of gravity will remain within limits during all phases of flight.
4. Describes the effects of various atmospheric conditions on the helicopter's performance.
5. Understands the cause and effects of retreating blade stall.
6. Considers circumstances when operating within "avoid areas" of the height/velocity diagram.
7. Is aware of situations that lead to loss of tail rotor/antitorque effectiveness (unanticipated yaw).
8. Determines whether the computed performance is within the helicopter's capabilities and operating limitations.

## F. TASK:   OPERATION OF SYSTEMS

REFERENCES:   AC 61-13; Helicopter Flight Manual.

**Objective.** To determine that the applicant exhibits knowledge of the elements related to the appropriate normal operating procedures and limitations of the following systems by explaining:

1. Primary flight controls, trim, and, if installed, stability control.
2. Powerplant.
3. Main rotor and antitorque.
4. Landing gear, brakes, steering, skids, or floats, as applicable.
5. Fuel, oil, and hydraulic.
6. Electrical.
7. Pitot-static, vacuum/pressure, and associated flight instruments, if applicable.
8. Environmental.
9. Anti-icing, including carburetor heat, if applicable.
10. Avionics equipment.

## G. TASK:   MINIMUM EQUIPMENT LIST
### Change 1

REFERENCE:    FAR Part 91.

**Objective.**  To determine that the applicant exhibits knowledge of the elements related to the FAA-approved minimum equipment list (MEL) by explaining:

1. Required instruments and equipment for day/night VFR.
2. Procedures and limitations for operating the aircraft with inoperative instruments and equipment with and without an MEL.
3. Process for obtaining an MEL to include a letter of authorization.
4. When a special flight permit would be required.
5. Procedures for obtaining a special flight permit.

## H. TASK:   AEROMEDICAL FACTORS

REFERENCES:  AC 61-21; AIM.

**Objective.**  To determine that the applicant exhibits knowledge of the elements related to aeromedical factors by explaining:

1. The symptoms, causes, effects, and corrective actions of at least three of the following—

   a. hypoxia.
   b. hyperventilation.
   c. middle ear and sinus problems.
   d. spatial disorientation.
   e. motion sickness.
   f. carbon monoxide poisoning.
   g. stress and fatigue.

2. The effects of alcohol and drugs, including over-the-counter drugs.
3. The effects of nitrogen excesses during scuba dives upon a pilot and/or passenger in flight.

## II. AREA OF OPERATION: PREFLIGHT PROCEDURES

### A. TASK: PREFLIGHT INSPECTION

REFERENCES: AC 61-13; Helicopter Flight Manual.

**Objective.** To determine that the applicant:

1. Exhibits knowledge of the elements related to a preflight inspection including which items must be inspected, for what reason, and how to detect possible defects.
2. Inspects the helicopter by systematically following a prescribed checklist.
3. Verifies that the helicopter is in condition for safe flight, notes any discrepancy, and determines if maintenance is required.

### B. TASK: COCKPIT MANAGEMENT

REFERENCES: FAR Part 91; AC 91-32; Helicopter Flight Manual.

**Objective.** To determine that the applicant:

1. Exhibits knowledge of the elements related to efficient cockpit management procedures, and related safety factors.
2. Organizes and arranges material and equipment in a manner that makes the items readily available.
3. Briefs the occupants on the use of safety belts, rotor blade avoidance, and emergency procedures.
4. Completes the prescribed checklist.

## C. TASK:   ENGINE STARTING AND ROTOR ENGAGEMENT

REFERENCES:   AC 61-13, AC 91-13, AC 91-42; Helicopter Flight Manual.

**Objective.**   To determine that the applicant:

1. Exhibits knowledge of the elements related to correct engine starting procedures, including the use of an external power source, starting under various atmospheric conditions, awareness of other persons and property during start, and the effects of using incorrect starting procedures.
2. Ensures proper rotor blade clearance, and frictions flight controls, as necessary.
3 Accomplishes correct starting procedures.
4. Prevents helicopter movement during and after the engine start.
5. Monitors engine instruments after start for proper engine RPM, temperature, and pressures.
6. Completes the prescribed checklist.

## D. TASK:   BEFORE TAKEOFF CHECK

REFERENCES:   AC 61-13; Helicopter Flight Manual.

**Objective.**   To determine that the applicant:

1. Exhibits knowledge of the elements related to the before takeoff check, including the reasons for checking the items and how to detect malfunctions.
2. Positions the helicopter properly considering other aircraft, surface conditions, and if applicable, existing wind conditions.
3. Properly divides attention.
4. Accomplishes the before takeoff check and ensures that the helicopter is in safe operating condition.
5. Reviews expected takeoff distance, if necessary.
6. Reviews takeoff emergency procedures.
7. Ensures no conflict with traffic prior to takeoff.
8. Ensures that the engine temperature and oil pressure are suitable, and rotor RPM is adequate for takeoff.
9. Completes the prescribed checklist, if applicable.

# III. AREA OF OPERATION: AIRPORT AND HELIPORT OPERATIONS

### A. TASK:  RADIO COMMUNICATIONS AND ATC LIGHT SIGNALS

REFERENCE:  AIM.

**Objective.** To determine that the applicant:

1. Exhibits knowledge of the elements related to radio communications, radio failure, and ATC light signals.
2. Selects appropriate frequencies for facilities to be used.
3. Transmits using recommended phraseology.
4. Acknowledges radio communications and complies with instructions.
5. Uses prescribed procedures following radio communications failure.
6. Interprets and complies with ATC light signals.

### B. TASK:  TRAFFIC PATTERNS

REFERENCES:  FAR Part 91; AC 61-21; AIM, Helicopter Flight Manual.

**Objective.** To determine that the applicant:

1. Exhibits knowledge of the elements related to traffic pattern procedures at each class airspace airport, runway incursion avoidance, collision and wake turbulence avoidance, and approach procedure when wind shear is reported.
2. Follows the established traffic pattern procedures, instructions, and rules.
3. Maintains proper spacing from other traffic or avoids the flow of fixed wing aircraft.
4. Remains aware of the possibility of wind shear and/or wake turbulence.
5. Maintains proper ground track with crosswind correction, if necessary.
6. Remains oriented with runway and/or landing area in use.
7. Maintains and holds traffic pattern altitude, ±100 feet (30 meters), and the appropriate airspeed, ±10 knots.
8. Completes the prescribed checklist, if applicable.

## C. TASK: AIRPORT AND HELIPORT MARKINGS AND LIGHTING

REFERENCE: AIM.

**Objective.** To determine that the applicant:

1. Exhibits knowledge of the elements related to airport and heliport markings and lighting.
2. Identifies and interprets airport and heliport markings and lighting.

# IV. AREA OF OPERATION: HOVERING MANEUVERS

## A. TASK: VERTICAL TAKEOFF AND LANDING

REFERENCES: AC 61-13; Helicopter Flight Manual.

**Objective.** To determine that the applicant:

1. Exhibits knowledge of the elements related to a vertical takeoff to a hover and landing from a hover.
2. Ascends to and maintains recommended hovering altitude, and descends from recommended hovering altitude in headwind, crosswind, and tailwind conditions.
3. Maintains RPM within normal limits.
4. Establishes recommended hovering altitude, ±1/2 of that altitude within 10 feet (3 meters) of the surface; if above 10 feet, ±5 feet (2 meters).
5. Avoids conditions that might lead to loss of tail rotor/antitorque effectiveness.
6. Keeps forward and sideward movement within 4 feet (1.2 meters) of a designated point, with no aft movement.
7. Descends vertically to within 4 feet (1.2 meters) of the designated touchdown point.
8. Maintains specified heading, ±10°.

## B. TASK: SLOPE OPERATIONS

REFERENCES: AC 61-13; Helicopter Flight Manual.

**Objective.** To determine that the applicant:

1. Exhibits knowledge of the elements related to slope operations.
2. Selects a suitable slope, approach, and direction considering wind effect, obstacles, dynamic rollover avoidance, and discharging passengers.
3. Properly moves toward the slope.
4. Maintains RPM within normal limits.
5. Makes a smooth positive descent to touch the upslope skid on the sloping surface.

6. Maintains positive control while lowering the downslope skid or landing gear to touchdown.
7. Recognizes when the slope is too steep and abandons the operation prior to reaching cyclic control stops.
8. Makes a smooth transition from the slope to a stabilized hover parallel to the slope.
9. Properly moves away from the slope.
10. Maintains the specified heading throughout the operation, ±10°.

## C. TASK: SURFACE TAXI

**NOTE:** This TASK applies to only helicopters equipped with wheel-type landing gear.

REFERENCES: AC 61-13; AIM, Helicopter Flight Manual.

**Objective.** To determine that the applicant:

1. Exhibits knowledge of the elements related to surface taxiing.
2. Surface taxies the helicopter from one point to another under headwind, crosswind, and tailwind conditions, with the landing gear in contact with the surface, avoiding conditions that might lead to loss of tail rotor/antitorque effectiveness.
3. Properly uses cyclic, collective, and brakes to control speed while taxiing.
4. Properly positions tailwheel, if applicable, locked or unlocked.
5. Maintains RPM within normal limits.
6. Maintains appropriate speed for existing conditions.
7. Stops helicopter within 4 feet (1.2 meters) of a specified point.
8. Maintains specified track within 4 feet (1.2 meters).

## D. TASK: HOVER TAXI

REFERENCES: AC 61-13; AIM, Helicopter Flight Manual.

**Objective.** To determine that the applicant:

1. Exhibits knowledge of the elements related to hover taxiing.
2. Hover taxies over specified ground references, demonstrating forward, sideward, and rearward hovering and hovering turns.
3. Maintains RPM within normal limits.
4. Maintains specified ground track within 4 feet (1.2 meters) on straight legs.
5. Maintains constant rate of turn at pivot points.
6. Maintains position within 4 feet (1.2 meters) of each pivot point during turns.
7. Makes 90°, 180°, and 360° pivoting turns, stopping within 10° of specified headings.
8. Maintains recommended hovering altitude, ±1/2 of that altitude within 10 feet (3 meters) of the surface, if above 10 feet, ±5 feet (2 meters).

## E. TASK: AIR TAXI

REFERENCES: AC 61-13; AIM, Helicopter Flight Manual.

**Objective.** To determine that the applicant:

1. Exhibits knowledge of the elements related to air taxiing.
2. Air taxies the helicopter from one point to another under headwind and crosswind conditions.
3. Maintains RPM within normal limits.
4. Selects a safe airspeed and altitude considering the possibility of an engine failure during taxi.
5. Maintains desired track and groundspeed in headwind and crosswind conditions, avoiding conditions that might lead to loss of tail rotor/antitorque effectiveness.
6. Maintains a specified altitude, ±10 feet (3 meters).

# V. AREA OF OPERATION:
## TAKEOFFS, LANDINGS, AND GO-AROUNDS

### A. TASK: NORMAL AND CROSSWIND TAKEOFF AND CLIMB

REFERENCES: AC 61-13; Helicopter Flight Manual.

**NOTE:** If a calm wind weather condition exists, the applicant's knowledge of the crosswind elements shall be evaluated through oral testing; otherwise a crosswind takeoff and climb shall be demonstrated.

**Objective.** To determine that the applicant:

1. Exhibits knowledge of the elements related to normal and crosswind takeoff and climb, including factors affecting performance, to include height/velocity information.
2. Establishes a stationary position on the surface or a stabilized hover, prior to takeoff in headwind and crosswind conditions.
3. Maintains RPM within normal limits.
4. Accelerates to manufacturer's recommended climb airspeed, ±10 knots.
5. Maintains proper ground track with crosswind correction, if necessary.
6. Remains aware of the possibility of wind shear and/or wake turbulence.
7. Completes the prescribed checklist, if applicable.

## B. TASK:   NORMAL AND CROSSWIND APPROACH

REFERENCES:   AC 61-13; Helicopter Flight Manual.

NOTE:   If a calm wind weather condition exists, the applicant's knowledge of the crosswind elements shall be evaluated through oral testing; otherwise a crosswind approach and landing shall be demonstrated.

**Objective.** To determine that the applicant:

1. Exhibits knowledge of the elements related to normal and crosswind approach.
2. Considers performance data, to include height/velocity information.
3. Considers the wind conditions, landing surface, and obstacles.
4. Selects a suitable termination point.
5. Establishes and maintains the recommended approach angle, and proper rate of closure.
6. Remains aware of the possibility of wind shear and/or wake turbulence.
7. Avoids situations that may result in settling-with-power.
8. Maintains proper ground track with crosswind correction, if necessary.
9. Arrives at the termination point, on the surface or at a stabilized hover, ±4 feet (1.2 meters).
10. Completes the prescribed checklist, if applicable.

## C. TASK: MAXIMUM PERFORMANCE TAKEOFF AND CLIMB

REFERENCES: AC 61-13; Helicopter Flight Manual.

**Objective.** To determine that the applicant:

1. Exhibits knowledge of the elements related to a maximum performance takeoff and climb.
2. Considers situations where this maneuver is recommended and factors related to takeoff and climb performance, to include height/velocity information.
3. Maintains RPM within normal limits.
4. Utilizes proper control technique to initiate takeoff and forward climb airspeed attitude.
5. Utilizes the maximum available takeoff power.
6. After clearing all obstacles, transitions to normal climb attitude, airspeed, ±10 knots, and power setting.
7. Remains aware of the possibility of wind shear and/or wake turbulence.
8. Maintains proper ground track with crosswind correction, if necessary.
9. Completes the prescribed checklist, if applicable.

## D. TASK: STEEP APPROACH

REFERENCES: AC 61-13; Helicopter Flight Manual.

**Objective.** To determine that the applicant:

1. Exhibits knowledge of the elements related to a steep approach.
2. Considers situations where this maneuver is recommended and factors related to a steep approach, to include height/velocity information.
3. Considers the wind conditions, landing surface, and obstacles.
4. Selects a suitable termination point.
5. Establishes and maintains the recommended approach angle, (15° maximum) and rate of closure.
6. Avoids situations that can result in settling-with-power.
7. Remains aware of the possibility of wind shear and/or wake turbulence.
8. Maintains proper ground track with crosswind correction, if necessary.
9. Arrives at the termination point, on the surface or at a stabilized hover, ±4 feet (1.2 meters).
10. Completes the prescribed checklist, if applicable.

## E. TASK: ROLLING TAKEOFF

NOTE: This TASK applies only to helicopters equipped with wheel-type landing gear.

REFERENCES: AC 61-13; Helicopter Flight Manual.

**Objective.** To determine that the applicant:

1. Exhibits knowledge of the elements related to a rolling takeoff.
2. Considers situations where this maneuver is recommended and factors related to takeoff and climb performance, to include height/velocity information.
3. Maintains RPM within normal limits.
4. Utilizes proper preparatory technique prior to initiating takeoff.
5. Initiates forward accelerating movement on the surface.
6. Transitions to a normal climb airspeed, ±10 knots, and power setting.
7. Remains aware of the possibility of wind shear and/or wake turbulence.
8. Maintains proper ground track with crosswind correction, if necessary.
9. Completes the prescribed checklist, if applicable.

## F. TASK:   SHALLOW APPROACH AND RUNNING/ROLL-ON LANDING

REFERENCES:   AC 61-13; Helicopter Flight Manual.

**Objective.**   To determine that the applicant:

1. Exhibits knowledge of the elements related to shallow approach and running/roll-on landing, including the purpose of the maneuver, factors affecting performance data, to include height/velocity information, and effect of landing surface texture.
2. Maintains RPM within normal limits.
3. Considers obstacles and other hazards.
4. Establishes and maintains the recommended approach angle, and proper rate of closure.
5. Remains aware of the possibility of wind shear and/or wake turbulence.
6. Maintains proper ground track with crosswind correction, if necessary.
7. Maintains a speed that will take advantage of effective translational lift during surface contact with landing gear parallel with the ground track.
8. Utilizes proper flight control technique after surface contact.
9. Completes the prescribed checklist, if applicable.

## G. TASK:   GO-AROUND

REFERENCES:   AC 61-13; Helicopter Flight Manual.

**Objective.**   To determine that the applicant:

1. Exhibits knowledge of the elements related to a go-around and when it is necessary.
2. Makes a timely decision to discontinue the approach to landing.
3. Maintains RPM within normal limits.
4. Establishes proper control input to stop descent and initiate climb.
5. Retracts the landing gear, if applicable, after a positive rate-of-climb indication.
6. Maintains proper ground track with crosswind correction, if necessary.
7. Transitions to a normal climb airspeed, ±10 knots.
8. Completes the prescribed checklist, if applicable.

# VI. AREA OF OPERATION: PERFORMANCE MANEUVERS

## A. TASK: RAPID DECELERATION

REFERENCES: AC 61-13; Helicopter Flight Manual.

**Objective.** To determine that the applicant:

1. Exhibits knowledge of the elements related to rapid deceleration.
2. Maintains RPM within normal limits.
3. Properly coordinates all controls throughout the execution of the maneuver.
4. Maintains an altitude that will permit safe clearance between the tail boom and the surface.
5. Decelerates and terminates in a stationary hover at the recommended hovering altitude.
6. Maintains heading throughout the maneuver, ±10°.

## B. TASK: STRAIGHT IN AUTOROTATION

REFERENCES: AC 61-13; Helicopter Flight Manual.

**Objective.** To determine that the applicant:

1. Exhibits knowledge of the elements related to a straight in autorotation terminating with a power recovery to a hover.
2. Selects a suitable touchdown area.
3. Initiates the maneuver at the proper point.
4. Establishes proper aircraft trim and autorotation airspeed, ±5 knots.
5. Maintains rotor RPM within normal limits.
6. Compensates for windspeed and direction as necessary to avoid undershooting or overshooting the selected landing area.
7. Utilizes proper deceleration, collective pitch application to a hover.
8. Comes to a hover within 100 feet (30 meters) of a designated point.

# VII. AREA OF OPERATION: NAVIGATION

## A. TASK: PILOTAGE AND DEAD RECKONING

REFERENCES: AC 61-21, AC 61-23, AC 61-84.

**Objective.** To determine that the applicant:

1. Exhibits knowledge of the elements related to pilotage and dead reckoning.
2. Correctly flies to at least the first planned checkpoint to demonstrate accuracy in computations.
3. Identifies and follows landmarks by relating the surface features to chart symbols.
4. Navigates by means of precomputed headings, groundspeed, and elapsed time.
5. Verifies the helicopter's position within 3 nautical miles (5.6 Km) of the flight planned route at all times.
6. Arrives at the en route checkpoints within 5 minutes of the ETA.
7. Corrects for, and records, the differences between preflight fuel, groundspeed, and heading calculations and those determined en route.
8. Maintains the appropriate altitude, ±200 feet (60 meters) and established heading, ±15°.

**B. TASK:  RADIO NAVIGATION AND RADAR SERVICES**

REFERENCES:  AC 61-21, AC 61-23, AC 61-84; Navigation Equipment Operation Manuals.

**Objective.**  To determine that the applicant:

1. Exhibits knowledge of the elements related to radio navigation and ATC radar services.
2. Selects and identifies the appropriate facilities or coordinates, as appropriate.
3. Locates the helicopter's position relative to the navigation facilities or coordinates, as appropriate.
4. Intercepts and tracks a given radial or bearing.
5. Locates position using cross radials, coordinates, or bearings.
6. Recognizes and describes the indication of station or way point passage.
7 Recognizes signal loss and takes appropriate action.
8. Uses proper communication procedures when utilizing ATC radar services.
9. Maintains the appropriate altitude, ±200 feet (60 meters).

**C. TASK:  DIVERSION**

REFERENCES:  AC 61-21, AC 61-23, AC 61-84.

**Objective.**  To determine that the applicant:

1. Exhibits knowledge of the elements related to procedures for diversion.
2. Selects an appropriate alternate airport or heliport and route.
3. Promptly diverts toward the alternate airport or heliport.
4. Makes an accurate estimate of heading, groundspeed, arrival time, and fuel consumption to the alternate airport or heliport.
5. Maintains the appropriate altitude, ±200 feet (60 meters) and established heading, ±15°.

# D. TASK: LOST PROCEDURES

REFERENCES: AC 61-21, AC 61-23, AC 61-84; AIM.

**Objective.** To determine that the applicant:

1. Exhibits knowledge of the elements related to lost procedures.
2. Selects the best course of action when given a lost situation.
3. Maintains the original or appropriate heading, and if necessary, climbs.
4. Attempts to identify nearest prominent landmark(s).
5. Uses available navigation aids and/or contacts an appropriate facility for assistance.
6. Plans a precautionary landing if deteriorating weather and/or fuel exhaustion is impending.

# VIII. AREA OF OPERATION: EMERGENCY OPERATIONS

**NOTE:** Tasks F through I are knowledge only TASKS.

## A. TASK: POWER FAILURE AT A HOVER

REFERENCES: AC 61-13; Helicopter Flight Manual.

**Objective.** To determine that the applicant:

1. Exhibits knowledge of the elements related to power failure at a hover.
2. Determines that the terrain below the aircraft is suitable for a safe touchdown.
3. Performs autorotation from a stationary or forward hover into the wind at recommended altitude, and RPM, while maintaining established heading, ±10°.
4. Touches down with minimum sideward movement, and no rearward movement.
5. Exhibits orientation, division of attention, and proper planning.

## B. TASK: POWER FAILURE AT ALTITUDE

REFERENCES: AC 61-13; Helicopter Flight Manual.

**NOTE:** Simulated power failure at altitude shall be given over areas where actual touchdowns can safely be completed in the event of an actual powerplant failure.

**Objective.** To determine that the applicant:

1. Exhibits knowledge of the elements related to power failure at altitude.
2. Establishes an autorotation and selects a suitable landing area.
3. Establishes proper aircraft trim and autorotation airspeed, ±5 knots.
4. Maintains rotor RPM within normal limits.
5. Compensates for windspeed and direction as necessary to avoid undershooting or overshooting the selected landing area.
6. Terminates approach with a power recovery at a safe altitude when directed by the examiner.

# C. TASK: SYSTEMS AND EQUIPMENT MALFUNCTIONS

REFERENCES: AC 61-13; Helicopter Flight Manual.

**Objective.** To determine that the applicant:

1. Exhibits knowledge of the elements related to causes, indications, and pilot actions for various systems and equipment malfunctions.
2. Analyzes the situation and takes action, appropriate to the helicopter used for the practical test, in at least three of the following areas—

   a. engine/oil and fuel.
   b. hydraulic, if applicable.
   c. electrical.
   d. carburetor or induction icing.
   e. smoke and/or fire.
   f. flight control/trim.
   g. pitot static/vacuum and associated flight instruments, if applicable.
   h. rotor and/or antitorque.
   i. various frequency vibrations and the possible components that may be affected.
   j. any other emergency unique to the helicopter flown.

# D. TASK: SETTLING-WITH-POWER

REFERENCES: AC 61-13; Helicopter Flight Manual.

**Objective.** To determine that the applicant:

1. Exhibits knowledge of the elements related to settling-with-power.
2. Selects an altitude that will allow recovery to be completed no less than 1,000 feet (300 meters) AGL or, if applicable, the manufacturer's recommended altitude, whichever is higher.
3. Promptly recognizes and announces the onset of settling-with-power.
4. Utilizes the appropriate recovery procedure.

**E. TASK:   LOW ROTOR RPM RECOVERY**

REFERENCES: AC 61-13; Appropriate Manufacturer's Safety Notices; Helicopter Flight Manual.

**NOTE:** The examiner may test the applicant orally on this TASK if helicopter used for the practical test has a governor that cannot be disabled.

**Objective.** To determine that the applicant:

1. Exhibits knowledge of the elements related to low rotor RPM recovery, including the combination of conditions that are likely to lead to this situation.
2. Detects the development of low rotor RPM and initiates prompt corrective action.
3. Utilizes the appropriate recovery procedure.

**F. TASK:   DYNAMIC ROLLOVER**

REFERENCES: AC 61-13, AC 90-87; Helicopter Flight Manual.

**Objective.** To determine that the applicant:

1. Exhibits knowledge of the elements related to the aerodynamics of dynamic rollover.
2. Understands the interaction between the antitorque thrust, crosswind, slope, CG, cyclic, and collective pitch control in contributing to dynamic rollover.
3. Explains preventive flight technique during takeoffs, landings, and slope operations.

**G. TASK:   GROUND RESONANCE**

REFERENCES: AC 61-13; Helicopter Flight Manual.

**Objective.** To determine that the applicant:

1. Exhibits knowledge of the elements related to a fully articulated rotor system and the aerodynamics of ground resonance.
2. Understands the conditions that contribute to ground resonance.
3. Explains preventive flight technique during takeoffs and landings.

## H. TASK: LOW G CONDITIONS

REFERENCE:    Helicopter Flight Manual.

**Objective.** To determine that the applicant:

1. Exhibits knowledge of the elements related to low G conditions.
2. Understands and recognizes the situations that contribute to low G conditions.
3. Explains proper recovery procedures.

## I. TASK: EMERGENCY EQUIPMENT AND SURVIVAL GEAR

REFERENCE:    Helicopter Flight Manual.

**Objective.** To determine that the applicant:

1. Exhibits knowledge of the elements related to emergency equipment appropriate to the helicopter used for the practical test by describing—

   a. purpose of such equipment.
   b. location in the helicopter.
   c. method of operation.
   d. servicing requirements.
   e. method of safe storage.

2. Exhibits knowledge of the elements related to survival gear by describing—

   a. survival gear appropriate for operation in various climatological and topographical environments.
   b. location in the helicopter.
   c. method of operation.
   d. servicing requirements.
   e. method of safe storage.

# IX. AREA OF OPERATION:
## NIGHT OPERATIONS

Change 2
5/21/97

## A. TASK: PHYSIOLOGICAL ASPECTS OF NIGHT FLYING

REFERENCES: AC 61-21; AIM.

**Objective.** To determine that the applicant exhibits knowledge of the elements related to the physiological aspects of night flying by explaining:

1. The function of various parts of the eye essential for night vision.
2. Adaptation of the eye to changing light.
3. Correct use of the eye to accommodate changing light.
4. Coping with illusions created by various light conditions.
5. Effects of the pilot's physical condition on visual acuity.
6. Methods for increasing vision effectiveness.

## B. TASK: LIGHTING AND EQUIPMENT FOR NIGHT FLYING

REFERENCES: AC 61-21, AC 61-23; Helicopter Flight Manual.

**Objective.** To determine that the applicant:

1. Exhibits knowledge of the elements related to lighting and equipment for night flying by explaining—

   a. the types and uses of various personal lighting devices.
   b. the required equipment, and location of external navigation lighting of the helicopter.
   c. the meaning of various airport, heliport and navigation lights, the method of determining their status, and the procedure for airborne activation of runway lights.

2. Locates and identifies switches, spare fuses, and circuit breakers pertinent to night operations.

# X. AREA OF OPERATION: POST-FLIGHT PROCEDURES

## TASK: AFTER LANDING AND SECURING

REFERENCES: AC 61-13; Helicopter Flight Manual.

**Objective.** To determine that the applicant:

1. Exhibits knowledge of the elements related to after-landing procedures, including local and ATC operations, ramp safety, parking hand signals, shutdown, securing, and post-flight inspection.
2. Minimizes the hazardous effects of rotor downwash during hovering.
3. Selects a suitable parking area while considering wind and the safety of nearby persons and property.
4. Follows the recommended procedure for engine shutdown, securing the cockpit, securing rotor blades, and discharging passengers.
5. Performs a satisfactory post-flight inspection.
6. Completes the prescribed checklist, if applicable.

# COMMERCIAL PILOT ROTORCRAFT/HELICOPTER

## Practical Test Standards

## Section 1

**1996**

# NOTE

Material in FAA-S-8081-16 will be effective April 1, 1996. This version contains change 1. All previous editions of the Commercial Pilot — Rotorcraft (Helicopter) Practical Test Standards will be obsolete as of this date.

# RECORD OF CHANGES

**Change 1: 3/1/96**
**Reason:**

**1.** Commercial Pilot-Rotorcraft Practical Test Prerequisites, (page xii). Delete item #2, and renumber the remaining prerequisites. There is no requirement for possessing an instrument rating when making application for initial Rotorcraft-Helicopter, certification.

# FOREWORD

The Commercial Pilot — Rotorcraft (Helicopter) Practical Test Standards (PTS) book has been published by the Federal Aviation Administration (FAA) to establish the standards for commercial pilot certification practical tests for the rotorcraft category, helicopter class. FAA inspectors and designated pilot examiners shall conduct practical tests in compliance with these standards. Flight instructors and applicants should find these standards helpful during training and when preparing for the practical test.

_____

William J. White
Deputy Director, Flight Standards Service

FAA-S-8081-16 **Commercial**

# CONTENTS

# INTRODUCTION

The Flight Standards Service of the Federal Aviation Administration (FAA) has developed this practical test book as a standard to be used by FAA inspectors and designated pilot examiners when conducting commercial pilot – rotorcraft (helicopter) practical tests. Flight instructors are expected to use this book when preparing applicants for practical tests. Applicants should be familiar with this book and refer to these standards during their training.

Information considered directive in nature is described in this practical test book in terms such as "shall" and "must" indicating the actions are mandatory. Guidance information is described in terms such as "should" and "may" indicating the actions are desirable or permissive but not mandatory.

The FAA gratefully acknowledges the valuable assistance provided by a nationwide public "Job Task Analysis" team that developed the knowledge, skills, and abilities that appear in this book. We would also like to thank the many individuals and organizations who contributed their time and talent in assisting with the revision of these practical test standards.

This publication may be obtained from FedWorld through the use of a computer modem or purchased from the Superintendent of Documents, U.S. Government Printing Office, Washington, DC 20402.

Comments regarding this publication should be sent to:

U.S. Department of Transportation
Federal Aviation Administration
Flight Standards Service
Operations Support Branch, AFS-630
P.O. Box 25082
Oklahoma City, OK 73125

## PRACTICAL TEST STANDARD CONCEPT

Federal Aviation Regulations (FAR's) specify the areas in which knowledge and skill must be demonstrated by the applicant before the issuance of a commercial pilot certificate. The FAR's provide the flexibility to permit the FAA to publish practical test standards containing specific TASKS in which pilot competency must be demonstrated. The FAA will revise this book whenever it is determined that changes are needed in the interest of safety. Adherence to the provisions of the regulations and the practical test standards is mandatory for the evaluation of commercial pilot applicants.

## COMMERCIAL PILOT ROTORCRAFT PRACTICAL TEST BOOK DESCRIPTION

This test book contains the following commercial pilot practical test standards:

Section 1    Rotorcraft, Helicopter — Commercial Pilot

The Commercial Pilot Rotorcraft Practical Test Standards include the AREAS OF OPERATION and TASKS for the issuance of an initial commercial pilot certificate and for the addition of category and/or class ratings to that certificate.

## PRACTICAL TEST STANDARD DESCRIPTION

AREAS OF OPERATION are phases of the practical test arranged in a logical sequence within this standard. They begin with preflight preparation and end with post-flight procedures. The examiner, however, may conduct the practical test in any sequence that results in a complete and efficient test.

The REFERENCE identifies the publication(s) that describe(s) the TASK. Descriptions of TASKS are not included in the standards because this information can be found in the reference list. Publications other than those listed may be used, if their content conveys substantially the same meaning as the listed publications.

Reference list:

| | |
|---|---|
| **FAR Part 43** | Maintenance, Preventive Maintenance, Rebuilding, and Alteration |
| **FAR Part 61** | Certification: Pilots and Flight Instructors |
| **FAR Part 67** | Medical Standards and Certification |
| **FAR Part 91** | General Operating and Flight Rules |

| | |
|---|---|
| **NTSB Part 830** | Notification and Reporting of Aircraft Accidents and Incidents |
| **AC 00-2** | Advisory Circular Checklist |
| **AC 00-6** | Aviation Weather |
| **AC 00-45** | Aviation Weather Services |
| **AC 61-13** | Basic Helicopter Handbook |
| **AC 61-21** | Flight Training Handbook |
| **AC 61-23** | Pilot's Handbook of Aeronautical Knowledge |
| **AC 61-65** | Certification: Pilots and Flight Instructors |
| **AC 61-84** | Role of Preflight Preparation |
| **AC 90-48** | Pilots' Role in Collision Avoidance |
| **AC 90-87** | Helicopter Dynamic Rollover |
| **AC 91-13** | Cold Weather Operation of Aircraft |
| **AC 91-23** | Pilot's Weight and Balance Handbook |
| **AC 91-32** | Safety In and Around Helicopters |
| **AC 91-42** | Hazards of Rotating Propeller and Helicopter Rotor Blades |
| **AIM** | Aeronautical Information Manual |
| **AFD** | Airport Facility Directory |
| **NOTAM's** | Notices to Airmen |
| | Helicopter Flight Manuals |
| | Industry Related Manuals |

The Objective lists the important elements that must be satisfactorily performed to demonstrate competency in a TASK. The Objective includes:

1. specifically what the applicant should be able to do;
2. the conditions under which the TASK is to be performed; and
3. the acceptable standards of performance.

## USE OF THE PRACTICAL TEST STANDARDS BOOK

The Commercial Pilot Rotorcraft Practical Test Standards have been designed to evaluate the competency of commercial pilots in both knowledge and skill. Commercial pilots are professionals engaged in various flight activities for compensation or hire. Because of their professional status, they should exhibit a significantly higher level of knowledge and skill than the private pilot. Although some TASKS listed are similar to those in the Private Pilot Rotorcraft Practical Test Standards, the wording used in the Commercial Pilot Rotorcraft Practical Test Standards reflects a higher level of competency expected of a commercial pilot applicant in performing these similar TASKS.

The FAA requires that all practical tests be conducted in accordance with the appropriate Commercial Pilot Practical Test Standards and the policies set forth in this INTRODUCTION. Commercial pilot applicants shall be evaluated in **ALL** TASKS included in the AREAS OF OPERATION of the appropriate practical test standard.

In preparation for the practical test, the examiner shall develop a written "plan of action." The "plan of action" shall include all TASKS in each AREA OF OPERATION.

The examiner is not required to follow the precise order in which the AREAS OF OPERATION and TASKS appear in this book. The examiner may change the sequence or combine TASKS with similar objectives to meet the orderly and efficient flow of the practical test. For example, lost procedures  may be combined with radio navigation. The examiner's "plan of action" shall include the order and combination of TASKS to be demonstrated by the applicant in a manner that will result in an efficient and valid test.

Examiners shall place special emphasis upon areas of aircraft operation that are most critical to flight safety. Among these areas are precise aircraft control and sound judgment in decision making. Although these areas may or may not be shown under each TASK, they are essential to flight safety and shall receive careful evaluation throughout the practical test. THE EXAMINER SHALL ALSO EMPHASIZE WAKE TURBULENCE AVOIDANCE, LOW LEVEL WIND SHEAR, INFLIGHT COLLISION AVOIDANCE, RUNWAY INCURSION AVOIDANCE, AND CHECKLIST USAGE.

The examiner is expected to use good judgment in the performance of simulated  emergency procedures.  The use of the safest means for simulation is expected. Consideration must always be given to local conditions (both meteorological and topographical), the examiner's level of performance at the time of the test, as well as the applicant's, ATC workload, and the relative condition of the aircraft used. If the procedure being evaluated would put the maneuver in jeopardy of safe operation, it is expected that the applicant shall simulate that portion of the maneuver, i.e.—engine governor, trim system malfunction, etc., unless otherwise indicated by the NOTE in a particular AREA OF OPERATION or TASK.

## COMMERCIAL PILOT ROTORCRAFT PRACTICAL TEST PREREQUISITES

An applicant for the commercial pilot rotorcraft practical test is required by Federal Aviation Regulations to:

1. possess a private pilot certificate with a helicopter or gyroplane rating, if a commercial pilot certificate with a helicopter or gyroplane rating is sought, or meet the flight experience required for a private pilot certificate and pass the private helicopter or gyroplane knowledge and practical test;
2. pass the appropriate commercial pilot knowledge test since the beginning of the 24th month before the month in which the practical test is taken;
3. obtain the applicable instruction and aeronautical experience prescribed for the commercial pilot certificate or training sought;
4. for initial certification, hold at least a current second-class medical certificate issued under FAR Part 67;
5. be at least 18 years of age, and;
6. obtain a written statement from an appropriately certificated flight instructor certifying that the applicant has been given flight instruction in preparation for the practical test within 60 days preceding the date of application. The statement shall also state that the instructor finds the applicant competent to pass the practical test and that the applicant has satisfactory knowledge of the subject area(s) in which a deficiency was indicated by the airman knowledge test report.

AC 61-65, Certification: Pilots and Flight Instructors, states that the instructor may sign the instructor's recommendation on the reverse side of FAA Form 8710-1, Airman Certificate and/or Rating Application, in lieu of the previous statement, provided all appropriate FAR Part 61 requirements are substantiated by reliable records.

# AIRCRAFT AND EQUIPMENT REQUIRED FOR THE PRACTICAL TEST

The commercial pilot applicant is required by FAR Section 61.45 to provide an airworthy, certificated aircraft for use during the practical test. This section further requires that the aircraft:

1. have fully functioning dual controls, except as provided in this FAR Section; and
2. be capable of performing **ALL** appropriate TASKS for the commercial pilot certificate and have no operating limitations that prohibit the performance of those TASKS.

## METRIC CONVERSION INITIATIVE

To assist the pilots in understanding and using the metric measurement system, the practical test standards refer to the metric equivalent of various altitudes throughout. The inclusion of meters is intended to familiarize pilots with its use. The metric altimeter is arranged in 10 meter increments; therefore, when converting from feet to meters, the exact conversion, being too exact for practical purposes, is rounded to the nearest 10 meter increment or even altitude as necessary.

## POSITIVE EXCHANGE OF FLIGHT CONTROLS

During the practical test, there must always be a clear understanding of who has control of the aircraft. Prior to the flight, a briefing should be conducted that includes the procedure for the exchange of flight controls. A positive three-step process in the exchange of flight controls between pilots is a proven procedure and one that is recommended.

When the examiner wishes to take the controls to allow the applicant to adjust the seat, headset, etc., he/she will say "I have the controls." The applicant will acknowledge immediately by saying, "You have the controls." The examiner again says, "I have the controls." When control is returned to the applicant, follow the same procedure. A visual check is recommended to verify that the exchange has occurred. There should never be any doubt as to who is flying the aircraft.

## USE OF DISTRACTIONS DURING PRACTICAL TESTS

Numerous studies indicate that many accidents have occurred when the pilot has been distracted during critical phases of flight. To evaluate the pilot's ability to utilize proper control technique while dividing attention both inside and/or outside the cockpit, the examiner shall cause a realistic distraction during the **flight** portion of the practical test to evaluate the applicant's ability to divide attention while maintaining safe flight.

## APPLICANT'S USE OF PRESCRIBED CHECKLISTS

Throughout the practical test, the applicant is evaluated on the use of the prescribed checklist. The situation may be such that the use of the checklist while accomplishing the elements of the objective would be either unsafe or impractical, especially in a single-pilot operation. In this case, it may be more prudent to review the checklist after the elements have been met.

## CREW RESOURCE MANAGEMENT (CRM)

CRM "...refers to the effective use of ALL available resources; human resources, hardware, and information." Human resources "...includes all other groups routinely working with the cockpit crew (or pilot) who are involved in decisions that are required to operate a flight safely. These groups include, but are not limited to: dispatchers, cabin crewmembers, maintenance personnel, and air traffic controllers." CRM is not a single TASK, it is a set of skill competencies that must be evident in all TASKS in this PTS as applied to either single pilot or a crew operation.

## FLIGHT INSTRUCTOR RESPONSIBILITY

An appropriately rated flight instructor is responsible for training the commercial pilot applicant to acceptable standards in **all** subject matter areas, procedures, and maneuvers included in the TASKS within the appropriate commercial pilot practical test standard. Because of the impact of their teaching activities in developing safe, proficient pilots, flight instructors should exhibit a high level of knowledge, skill, and ability. Additionally, the flight instructor must certify that the applicant is able to perform safely as a commercial pilot and is competent to pass the required practical test.

Throughout the applicant's training, the flight instructor is responsible for emphasizing the performance of effective visual scanning, collision avoidance, and runway incursion avoidance procedures.

# EXAMINER[1] RESPONSIBILITY

The examiner conducting the practical test is responsible for determining that the applicant meets the acceptable standards of knowledge and skill of each TASK within the appropriate practical test standard. Since there is no formal division between the **oral** and **skill** portions of the practical test, this becomes an ongoing process throughout the test. To avoid unnecessary distractions, oral questioning should be used judiciously at all times, especially during the flight portion of the practical test.

Examiners shall test to the greatest extent practicable the applicant's correlative abilities rather than mere rote enumeration of facts throughout the practical test.

Throughout the flight portion of the practical test, the examiner shall evaluate the applicant's use of visual scanning and collision avoidance procedures.

## SATISFACTORY PERFORMANCE

Satisfactory performance to meet the requirements for certification is based on the applicant's ability to safely:

1.  perform the approved areas of operation for the certificate or rating sought within the approved standards;
2.  demonstrate mastery of the aircraft with the successful outcome of each task performed never seriously in doubt;
3.  demonstrate sound judgment aeronautical decision making and skilled compentencies in CRM.

## UNSATISFACTORY PERFORMANCE

If, in the judgment of the examiner, the applicant does not meet the standards of performance of any TASK performed, the associated AREA OF OPERATION is failed and therefore, the practical test is failed. The examiner or applicant may discontinue the test any time after the failure of an AREA OF OPERATION makes the applicant ineligible for the certificate or rating sought. The test will be continued[1] ONLY with the consent of the applicant. If the test is either continued or discontinued, the applicant is entitled credit for only those TASKS satisfactorily performed. However, during the retest and at the discretion of the examiner, any TASK may be re-evaluated including those previously passed.

---

[1] The word "examiner" denotes either the FAA inspector or FAA designated pilot examiner who conducts the practical test.

Typical areas of unsatisfactory performance and grounds for disqualification are:

1. Any action or lack of action by the applicant that requires corrective intervention by the examiner to maintain safe flight.
2. Failure to use proper and effective visual scanning techniques to clear the area before and while performing maneuvers.
3. Consistently exceeding tolerances stated in the Objectives.
4. Failure to take prompt corrective action when tolerances are exceeded.

When a disapproval notice is issued, the examiner will record the applicant's unsatisfactory performance and tasks not completed in terms of AREA OF OPERATIONS appropriate to the practical test conducted.

## USE OF RATING TASKS TABLES

If an applicant already holds a commercial pilot certificate, use the appropriate table at the beginning of each section, to determine which TASKS are required on the practical test. However, at the discretion of the examiner, the applicant's competence in any TASK may be evaluated, if indications of the applicant's performance suggests that such action is appropriate.

If the applicant holds more than one category or class rating at the commercial level, and the table indicates differing required TASKS, the "least restrictive" entry applies. For example, if "ALL" and "NONE" are indicated for one AREA OF OPERATION, the "NONE" entry applies. If "B" and "B, C" are indicated, the "B" entry applies.

# CONTENTS: SECTION 1

## Addition of a Rotorcraft/Helicopter rating to an existing Commercial Pilot Certificate

| Area of Operation | Required TASKS are indicated by either the TASK letter(s) that apply(s) or an indication that all or none of the TASKS must be tested. | | | | | | | | |
|---|---|---|---|---|---|---|---|---|---|
| | **COMMERCIAL PILOT RATING(S) HELD** | | | | | | | | |
| | ASEL | ASES | AMEL | AMES | RG | Non-Power Glider | Power Glider | Free Balloon | Airship |
| I | E,F,G | E,F,G | E,F,G | E,F,G | E,F,G | E,F,G, I,J | E,F,G, I,J | E,F,G, I,J | E,F,G |
| II | ALL | ALL | ALL | ALL | ALL | ALL | ALL | ALL | ALL |
| III | B,C | B,C | B,C | B,C | ALL | ALL | ALL | ALL | B,C |
| IV | ALL | ALL | ALL | ALL | ALL | ALL | ALL | ALL | ALL |
| V | ALL | ALL | ALL | ALL | ALL | ALL | ALL | ALL | ALL |
| VI | ALL | ALL | ALL | ALL | ALL | ALL | ALL | ALL | ALL |
| VII | NONE | NONE | NONE | NONE | B | B,C,D | B,C,D | B,C,D | NONE |
| VIII | ALL | ALL | ALL | ALL | ALL | ALL | ALL | ALL | ALL |
| IX | ALL | ALL | ALL | ALL | ALL | ALL | ALL | ALL | ALL |
| X | ALL | ALL | ALL | ALL | ALL | ALL | ALL | ALL | ALL |

## Appendix 1

### TASK VS. SIMULATION DEVICE CREDIT

Examiners conducting the Commercial Pilot—Helicopter Practical Tests with simulation devices should consult appropriate documentation to ensure that the device has been approved for training. The documentation for each device should reflect that the following activities have occurred:

1. The device must be evaluated, determined to meet the appropriate standards, and assigned the appropriate qualification level by the National Simulator Program Manager. The device must continue to meet qualification standards through continuing evaluations as outlined in the appropriate advisory circular (AC). For helicopter simulators, AC 120-63 (as amended), Helicopter Simulator Qualification, will be used.
2. The FAA must approve the device for specific TASKS.
3. The device must continue to support the level of student or applicant performance required by this PTS.

NOTE: Users of the following chart are cautioned that use of the chart alone is incomplete. The description and objective of each task as listed in the body of the PTS, including all notes, must also be incorporated for accurate simulation device use.

### USE OF CHART

X    Creditable.

X1   Creditable only if accomplished in conjunction with a running takeoff or running landing, as appropriate.

NOTE:  1. The helicopter may be used for all tasks.
2. Level C simulators may be used as indicated only if the applicant meets established pre-requisite experience requirements.
3. Level A helicopter simulator standards have not been defined.
4. Helicopter flight training devices have not been defined.

# FLIGHT TASK
## Areas of Operation:

# FLIGHT SIMULATION DEVICE LEVEL
| | 1 | 2 | 3 | 4 | 5 | 6 | 7 | A | B | C | D |
|---|---|---|---|---|---|---|---|---|---|---|---|

**II. Preflight Procedures**
- A. Preflight Inspection (Cockpit Only)
- B. Cockpit Management
- C. Engine Starting and Rotor Engagement (If applicable)
- D. Before Takeoff Check

**III. Airport and Heliport Operations**
- A. Radio Communications and ATC Light Signals
- B. Traffic Patterns
- C. Airport and Heliport Markings and Lighting

**IV. Hovering Maneuvers**
- A. Vertical Takeoff and Landing
- B. Slope Operations
- C. Surface Taxi
- D. Hover Taxi
- E. Air Taxi

**V. Takeoffs, Landings, and Go-Arounds**
- A. Normal and Crosswind Takeoff and Climb
- B. Normal and Crosswind Approach
- C. Maximum Performance Takeoff and Climb
- D. Steep Approach
- E. Rolling Takeoff
- F. Shallow Approach and Running /Roll-On
- G. Go-Around

| | 1 | 2 | 3 | 4 | 5 | 6 | 7 | A | B | C | D |
|---|---|---|---|---|---|---|---|---|---|---|---|
| **VI. Performance Maneuvers** | | | | | | | | | | | |
| A. Rapid Deceleration | | | | | | | | | | | |
| B. 180° Autorotations | | | | | | | | | | | |
| **VII. Navigation** | | | | | | | | | | | |
| A. Pilotage and Dead Reckoning | | | | | | | | | | X | X |
| B. Radio Navigation and Radar Services | | | | | | | | | | X | X |
| C. Diversion | | | | | | | | | | X | X |
| D. Lost Procedures | | | | | | | | | | X | X |
| **VIII. Emergency Operations** | | | | | | | | | | | |
| A. Power Failure at a Hover | | | | | | | | | | X | X |
| B. Power Failure at Altitude | | | | | | | | | | X | X |
| C. Systems and Equipment Malfunctions | | | | | | | | | | X | X |
| D. Settling-With-Power | | | | | | | | | | X | X |
| E. Low Rotor RPM Recovery | | | | | | | | | | X | X |
| F. Dynamic Roll Over | | | | | | | | | | | |
| G. Ground Resonance | | | | | | | | | | | |
| H. Low G Conditions | | | | | | | | | | | |
| I. Emergency Equipment and Survival Gear | | | | | | | | | | | |
| **IX. Special Operations** | | | | | | | | | | | |
| A. Confined Area Operations | | | | | | | | | | | |
| B. Pinnacle/Platform Operations | | | | | | | | | | | |
| **X. Postflight Procedures** | | | | | | | | | | | |
| A. After Landing and Securing | | | | | | | | | | | |

# APPLICANT'S PRACTICAL TEST CHECKLIST
## (HELICOPTER)
## APPOINTMENT WITH EXAMINER:

**EXAMINER'S NAME**_____

**LOCATION** _____

**DATE/TIME** _____

## ACCEPTABLE AIRCRAFT

- ☐ Aircraft Documents:
- ☐ Airworthiness Certificate
    Registration Certificate
    Operating Limitations
- ☐ Aircraft Maintenance Records:
    Logbook Record of Airworthiness Inspections
    and AD Compliance
- ☐ Pilot's Operating Handbook, FAA-Approved
    Helicopter Flight Manual
- ☐ FCC Station License

## PERSONAL EQUIPMENT

- ☐ View-Limiting Device
- ☐ Current Aeronautical Charts
- ☐ Computer and Plotter
- ☐ Flight Plan Form
- ☐ Flight Logs
- ☐ Current AIM, Airport Facility Directory, and Appropriate
    Publications

## PERSONAL RECORDS

- ☐ Identification—Photo/Signature ID
- ☐ Pilot Certificate
- ☐ Current and Appropriate Medical Certificate
- ☐ Completed FAA Form 8710-1, Airman Certificate and/or
    Rating Application with Instructor's Signature (if applicable)
- ☐ AC Form 8080-2, Airman Written Test Report, or
    Computer Test Report
- ☐ Pilot Logbook with Appropriate Instructor Endorsements
- ☐ FAA Form 8060-5, Notice of Disapproval (if applicable)
- ☐ Approved School Graduation Certificate (if applicable)
- ☐ Examiner's Fee (if applicable)

# EXAMINER'S PRACTICAL TEST CHECKLIST

## (HELICOPTER)

**APPLICANT'S NAME**_____

**LOCATION**_____

**DATE/TIME**_____

### I.   PREFLIGHT PREPARATION

- ☐ **A.** CERTIFICATES AND DOCUMENTS
- ☐ **B.** WEATHER INFORMATION
- ☐ **C.** CROSS-COUNTRY FLIGHT PLANNING
- ☐ **D.** NATIONAL AIRSPACE SYSTEM
- ☐ **E.** PERFORMANCE AND LIMITATIONS
- ☐ **F.** OPERATION OF SYSTEMS
- ☐ **G.** MINIMUM EQUIPMENT LIST
- ☐ **H.** AEROMEDICAL FACTORS
- ☐ **I.** PHYSIOLOGICAL ASPECTS OF NIGHT FLYING
- ☐ **J.** LIGHTING AND EQUIPMENT FOR NIGHT FLYING

### II.  PREFLIGHT PROCEDURES

- ☐ **A.** PREFLIGHT INSPECTION
- ☐ **B.** COCKPIT MANAGEMENT
- ☐ **C.** ENGINE STARTING AND ROTOR ENGAGEMENT
- ☐ **D.** BEFORE TAKEOFF CHECK

### III. AIRPORT AND HELIPORT OPERATIONS

- ☐ **A.** RADIO COMMUNICATIONS AND ATC LIGHT SIGNALS
- ☐ **B.** TRAFFIC PATTERNS
- ☐ **C.** AIRPORT AND HELIPORT MARKINGS AND LIGHTING

### IV. HOVERING MANEUVERS

- ☐ **A.** VERTICAL TAKEOFF AND LANDING
- ☐ **B.** SLOPE OPERATIONS
- ☐ **C.** SURFACE TAXI
- ☐ **D.** HOVER TAXI
- ☐ **E.** AIR TAXI

## V. TAKEOFFS, LANDINGS, AND GO-AROUNDS

- ☐ **A.** NORMAL AND CROSSWIND TAKEOFF AND CLIMB
- ☐ **B.** NORMAL AND CROSSWIND APPROACH
- ☐ **C.** MAXIMUM PERFORMANCE TAKEOFF AND CLIMB
- ☐ **D.** STEEP APPROACH
- ☐ **E.** ROLLING TAKEOFF
- ☐ **F.** SHALLOW APPROACH AND RUNNING/ROLL-ON LANDING
- ☐ **G.** GO-AROUND

## VI. PERFORMANCE MANEUVERS

- ☐ **A.** RAPID DECELERATION
- ☐ **B.** 180° AUTOROTATION

## VII. NAVIGATION

- ☐ **A.** PILOTAGE AND DEAD RECKONING
- ☐ **B.** RADIO NAVIGATION AND RADAR SERVICES
- ☐ **C.** DIVERSION
- ☐ **D.** LOST PROCEDURES

## VIII. EMERGENCY OPERATIONS

- ☐ **A.** POWER FAILURE AT A HOVER
- ☐ **B.** POWER FAILURE AT ALTITUDE
- ☐ **C.** SYSTEMS AND EQUIPMENT MALFUNCTIONS
- ☐ **D.** SETTLING-WITH-POWER
- ☐ **E.** LOW ROTOR RPM RECOVERY
- ☐ **F.** DYNAMIC ROLLOVER
- ☐ **G.** GROUND RESONANCE
- ☐ **H.** LOW G CONDITIONS
- ☐ **I.** EMERGENCY EQUIPMENT AND SURVIVAL GEAR

## IX. SPECIAL OPERATIONS

- ☐ **A.** CONFINED AREA OPERATION
- ☐ **B.** PINNACLE/PLATFORM OPERATIONS

## X. POST-FLIGHT PROCEDURES

- ☐ AFTER LANDING AND SECURING

# I. AREA OF OPERATION: PREFLIGHT PREPARATION

## A. TASK: CERTIFICATES AND DOCUMENTS

REFERENCES: FAR Parts 43, 61, 67, 91; AC 61-13, AC 61-23; Helicopter Flight Manual.

**Objective.** To determine that the applicant:

1. Exhibits knowledge of the elements related to certificates and documents by explaining—

    a. pilot certificate privileges and limitations.
    b. medical certificate class and duration.
    c. pilot logbook or flight records.

2. Exhibits knowledge of the elements related to certificates and documents by locating and explaining—

    a. airworthiness and registration certificates.
    b. operating limitations, placards, and instrument markings.
    c. weight and balance data and equipment list.
    d. airworthiness directives, compliance records, maintenance requirements, and appropriate records.

## B. TASK: WEATHER INFORMATION

REFERENCES: AC 00-6, AC 00-45, AC 61-23, AC 61-84; AIM.

**Objective.** To determine that the applicant:

1. Exhibits knowledge of the elements related to weather information from various sources with emphasis on—

    a. PIREP's.
    b. SIGMET's and AIRMET's.
    c. wind shear reports.
    d. use of weather reports and forecasts.

2. Makes a competent "go/no-go" decision based on available weather information.

# C. TASK: CROSS-COUNTRY FLIGHT PLANNING

REFERENCES: AC 61-21, AC 61-23, AC 61-84; Navigation Charts; Airport/Facility Directory; NOTAM's; AIM.

NOTE: In-flight demonstration of cross-country procedures by the applicant is tested under the area of operation NAVIGATION.

Objective. To determine that the applicant:

1. Exhibits knowledge of the elements related to cross-country flight planning by presenting and explaining a pre-planned VFR cross-country flight, as previously assigned by the examiner. The flight plan shall be to the first fuel stop necessary, based on maximum allowable passenger, baggage, and/or cargo loads using real-time weather.

2. Uses appropriate and current aeronautical charts.

3. Properly identifies airspace, obstacles, and terrain features, including discussion of wire strike avoidance techniques.

4. Selects easily identifiable en route checkpoints.

5. Selects most favorable altitudes, considering weather conditions and equipment capabilities.

6 Computes headings, flight time, and fuel requirements.

7. Selects appropriate navigation systems/facilities and communication frequencies.

8. Extracts and applies pertinent information from NOTAM's, Airport/Facility Directory, and other flight publications.

9. Completes a navigation log and simulates filing a VFR flight plan.

## D. TASK:   NATIONAL AIRSPACE SYSTEM

REFERENCES:  FAR Part 91; AIM.

**Objective.** To determine that the applicant exhibits knowledge of the elements related to the National Airspace System by explaining:

1. Basic VFR Weather Minimums – for all classes of airspace.
2. Airspace classes – their boundaries, pilot certification and helicopter equipment requirements for the following—

   a. Class A.
   b. Class B.
   c. Class C.
   d. Class D.
   e. Class E.
   f.  Class G.

3. Special use airspace and other airspace areas.

## E. TASK:   PERFORMANCE AND LIMITATIONS

REFERENCES:  AC 61-13, AC 61-84, AC 91-23; Helicopter Flight Manual.

**Objective.** To determine that the applicant:

1. Exhibits knowledge of the elements related to performance and limitations by explaining the use of charts, tables, and data to determine performance and the adverse effects of exceeding limitations.
2. Computes weight and balance, including adding, removing, and shifting weight.
3. Determines if the weight and center of gravity will remain within limits during all phases of flight.
4. Describes the effects of various atmospheric conditions on the helicopter's performance.
5. Understands the cause and effects of retreating blade stall.
6. Considers circumstances when operating within "avoid areas" of the height/velocity diagram.
7. Is aware of situations that lead to loss of tail rotor/antitorque effectiveness (unanticipated yaw).
8. Determines whether the computed performance is within the helicopter's capabilities and operating limitations.

# F. TASK: OPERATION OF SYSTEMS

REFERENCES: AC 61-13; Helicopter Flight Manual.

**Objective.** To determine that the applicant exhibits knowledge of the elements related to the appropriate normal operating procedures and limitations of the following systems by explaining:

1. Primary flight controls, trim, and, if installed, stability control.
2. Powerplant.
3. Main rotor and antitorque.
4. Landing gear, brakes, steering, skids, or floats, as applicable.
5. Fuel, oil, and hydraulic.
6. Electrical.
7. Pitot-static, vacuum/pressure and associated flight instruments, if applicable.
8. Environmental.
9. Anti-icing, including carburetor heat, if applicable.
10. Avionics equipment.

# G. TASK: MINIMUM EQUIPMENT LIST

REFERENCE: FAR Part 91.

**Objective.** To determine that the applicant exhibits knowledge of the elements related to the FAA-approved minimum equipment list (MEL) by explaining:

1. Which aircraft require the use of an MEL.
2. Airworthiness limitations imposed on aircraft operations with inoperative instruments or equipment.
3. Requirements for a letter of authorization from the FAA Flight Standards District Office.
4. Supplemental type certificates related to MEL's.
5. Instrument and equipment exceptions.
6. When a special flight permit would be required.
7. Procedures for obtaining a special flight permit.

## H. TASK:  AEROMEDICAL FACTORS

REFERENCES: AC 61-21; AIM.

**Objective.** To determine that the applicant exhibits knowledge of the elements related to aeromedical factors by explaining:

1. The symptoms, causes, effects, and corrective actions of at least four of the following—

   a. hypoxia.
   b. hyperventilation.
   c. middle ear and sinus problems.
   d. spatial disorientation.
   e. motion sickness.
   f. carbon monoxide poisoning.
   g. stress and fatigue.

2. The effects of alcohol and drugs, including over-the-counter drugs.
3. The effects of nitrogen excesses during scuba dives upon a pilot and/or passenger in flight.

## I. TASK:  PHYSIOLOGICAL ASPECTS OF NIGHT FLYING

REFERENCES: AC 61-21; AIM.

**Objective.** To determine that the applicant exhibits knowledge of the elements related to the physiological aspects of night flying by explaining:

1. The function of various parts of the eye essential for night vision.
2. Adaptation of the eye to changing light.
3. Correct use of the eye to accommodate changing light.
4. Coping with illusions created by various light conditions.
5. Effects of the pilot's physical condition on visual acuity.
6. Methods for increasing vision effectiveness.

## J. TASK: LIGHTING AND EQUIPMENT FOR NIGHT FLYING

REFERENCES: AC 61-21, AC 61-23; Helicopter Flight Manual.

**Objective.** To determine that the applicant:

1. Exhibits knowledge of the elements related to lighting and equipment for night flying by explaining—

   a. the types and uses of various personal lighting devices.
   b. the required equipment, and location of external navigation lighting of the helicopter.
   c. the meaning of various airport, heliport, and navigation lights, the method of determining their status, and the procedure for airborne activation of runway lights.

2. Locates and identifies switches, spare fuses, and circuit breakers pertinent to night operations.

## II. AREA OF OPERATION: PREFLIGHT PROCEDURES

### A. TASK:  PREFLIGHT INSPECTION

REFERENCES:  AC 61-13; Helicopter Flight Manual.

**Objective.** To determine that the applicant:

1. Exhibits knowledge of the elements related to a preflight inspection including which items must be inspected, for what reason, and how to detect possible defects.
2. Inspects the helicopter by systematically following a prescribed checklist.
3. Verifies that the helicopter is in condition for safe flight, notes any discrepancy, and determines if maintenance is required.

### B. TASK:  COCKPIT MANAGEMENT

REFERENCES:  FAR Part 91; AC 91-32; Helicopter Flight Manual.

**Objective.** To determine that the applicant:

1. Exhibits knowledge of the elements related to efficient cockpit management procedures, and related safety factors.
2. Organizes and arranges material and equipment in a manner that makes the items readily available.
3. Briefs or causes the briefing of occupants on the use of safety belts, rotor blade avoidance, and emergency procedures.
4. If applicable, briefs crew appropriately.
5. Completes the prescribed checklist.

## C. TASK: ENGINE STARTING AND ROTOR ENGAGEMENT

REFERENCES: AC 61-13, AC 91-13, AC 91-42; Helicopter Flight Manual.

**Objective.** To determine that the applicant:

1. Exhibits knowledge of the elements related to correct engine starting procedures, including the use of an external power source, starting under various atmospheric conditions, awareness of other persons and property during start, and the effects of using incorrect starting procedures.
2. Ensures proper rotor blade clearance, and frictions flight controls, as necessary.
3 Accomplishes correct starting procedures.
4. Prevents helicopter movement during and after the engine start.
5. Monitors engine instruments after start for proper engine RPM, temperature, and pressures.
6. Completes the prescribed checklist.

## D. TASK: BEFORE TAKEOFF CHECK

REFERENCES: AC 61-13; Helicopter Flight Manual.

**Objective.** To determine that the applicant:

1. Exhibits knowledge of the elements related to the before takeoff check, including the reasons for checking the items and how to detect malfunctions.
2. Positions the helicopter properly considering other aircraft, surface conditions, and if applicable, existing wind conditions.
3. Properly divides attention.
4. Accomplishes the before takeoff check and ensures that the helicopter is in safe operating condition.
5. Reviews expected takeoff distance, if necessary.
6. Reviews takeoff emergency procedures, and if applicable, briefs crew on procedures.
7. Ensures no conflict with traffic prior to takeoff.
8. Ensures that the engine temperature and oil pressure are suitable, and rotor RPM is adequate for takeoff.
9. Completes the prescribed checklist, if applicable.

*FAA-S-8081-16* **Commercial**

## III. AREA OF OPERATION:
## AIRPORT AND HELIPORT OPERATIONS

### A. TASK: RADIO COMMUNICATIONS AND ATC LIGHT SIGNALS

REFERENCE: AIM.

**Objective.** To determine that the applicant:

1. Exhibits knowledge of the elements related to radio communications, radio failure, and ATC light signals.
2. Selects appropriate frequencies for facilities to be used.
3. Transmits using recommended phraseology.
4. Acknowledges radio communications and complies with instructions.
5. Uses prescribed procedures following radio communications failure.
6. Interprets and complies with ATC light signals.

### B. TASK: TRAFFIC PATTERNS

REFERENCES: FAR Part 91; AC 61-21; AIM, Helicopter Flight Manual.

**Objective.** To determine that the applicant:

1. Exhibits knowledge of the elements related to traffic pattern procedures at each class airspace airport, runway incursion avoidance, collision and wake turbulence avoidance, and approach procedure when wind shear is reported.
2. Follows the established traffic pattern procedures, instructions, and rules.
3. Maintains proper spacing from other traffic or avoids the flow of fixed wing aircraft.
4. Remains aware of the possibility of wind shear and/or wake turbulence.
5. Maintains proper ground track with crosswind correction, if necessary.
6. Remains oriented with runway and/or landing area in use.
7. Maintains and holds traffic pattern altitude ±100 feet (30 meters), and appropriate airspeed, ±10 knots.
8. Completes the prescribed checklist, if applicable.

## C. TASK: AIRPORT AND HELIPORT MARKINGS AND LIGHTING

REFERENCE: AIM.

**Objective.** To determine that the applicant:

1. Exhibits knowledge of the elements related to airport and heliport markings and lighting.
2. Identifies and interprets airport and heliport markings and lighting.

# IV. AREA OF OPERATION: HOVERING MANEUVERS

## A. TASK: VERTICAL TAKEOFF AND LANDING

REFERENCES: AC 61-13; Helicopter Flight Manual.

**Objective.** To determine that the applicant:

1. Exhibits knowledge of the elements related to a vertical takeoff to a hover and landing from a hover.
2. Ascends to and maintains recommended hovering altitude, and descends from recommended hovering altitude in headwind, crosswind, and tailwind conditions.
3. Maintains RPM within normal limits.
4. Establishes recommended hovering altitude, ±1/2 of that altitude within 10 feet (3 meters) of the surface; if above 10 feet, ±5 feet (2 meters).
5. Avoids conditions that might lead to loss of tail rotor/antitorque effectiveness.
6. Keeps forward and sideward movement within 2 feet (.6 meters) of a designated point, with no aft movement.
7. Descends vertically to within 2 feet (.6 meters) of the designated touchdown point.
8. Maintains specified heading, ±10°.

## B. TASK: SLOPE OPERATIONS

REFERENCES: AC 61-13; Helicopter Flight Manual.

**Objective.** To determine that the applicant:

1. Exhibits knowledge of the elements related to slope operations.
2. Selects a suitable slope, approach, and direction considering wind effect, obstacles, dynamic rollover avoidance, and discharging passengers.
3. Properly moves toward the slope.
4. Maintains RPM within normal limits.
5. Makes a smooth positive descent to touch the upslope skid on the sloping surface.
6. Maintains positive control while lowering the downslope skid or landing gear to touchdown.
7. Recognizes when the slope is too steep and abandons the operation prior to reaching cyclic control stops.
8. Makes a smooth transition from the slope to a stabilized hover parallel to the slope.
9. Properly moves away from the slope.
10. Maintains the specified heading throughout the operation, ±5°.

## C. TASK: SURFACE TAXI

REFERENCES: AC 61-13; AIM, Helicopter Flight Manual.

**NOTE:** This TASK applies to only helicopters equipped with wheel-type landing gear.

**Objective.** To determine that the applicant:

1. Exhibits knowledge of the elements related to surface taxiing.
2. Surface taxies the helicopter from one point to another under headwind, crosswind, and tailwind conditions, with the landing gear in contact with the surface, avoiding conditions that might lead to loss of tail rotor/antitorque effectiveness.
3. Properly uses cyclic, collective, and brakes to control speed while taxiing.
4. Properly positions tailwheel, if applicable, locked or unlocked.
5. Maintains RPM within normal limits.
6. Maintains appropriate speed for existing conditions.
7. Stops helicopter within 2 feet (.6 meters) of a specified point.
8. Maintains specified track within 2 feet (.6 meters).

## D. TASK: HOVER TAXI

REFERENCES: AC 61-13; AIM, Helicopter Flight Manual.

**Objective.** To determine that the applicant:

1. Exhibits knowledge of the elements related to hover taxiing.
2. Hover taxies over specified ground references, demonstrating forward, sideward, and rearward hovering and hovering turns.
3. Maintains RPM within normal limits.
4. Maintains specified ground track within 2 feet (.6 meters) on straight legs.
5. Maintains constant rate of turn at pivot points.
6. Maintains position within 2 feet (.6 meters) of each pivot point during turns.
7. Makes 90°, 180°, and 360° pivoting turns, stopping within 10° of specified headings.
8. Maintains recommended hovering altitude, ±1/2 of that altitude within 10 feet (3 meters) of the surface, if above 10 feet, ±5 feet (2 meters).

*FAA-S-8081-16* **Commercial**

## E. TASK:  AIR TAXI

REFERENCES:  AC 61-13; AIM, Helicopter Flight Manual.

**Objective.**  To determine that the applicant:

1. Exhibits knowledge of the elements related to air taxiing.
2. Air taxies the helicopter from one point to another under headwind and crosswind conditions.
3. Maintains RPM within normal limits.
4. Selects a safe airspeed and altitude considering the possibility of an engine failure during taxi.
5. Maintains desired track and groundspeed in headwind and crosswind conditions, avoiding conditions that might lead to loss of tail rotor/antitorque effectiveness.
6. Maintains a specified altitude, ±5 feet (2 meters).

# V. AREA OF OPERATION:
## TAKEOFFS, LANDINGS, AND GO-AROUNDS

### A. TASK: NORMAL AND CROSSWIND TAKEOFF AND CLIMB

REFERENCES: AC 61-13; Helicopter Flight Manual.

**NOTE:** If a calm wind weather condition exists, the applicant's knowledge of the crosswind elements shall be evaluated through oral testing; otherwise a crosswind takeoff and climb shall be demonstrated.

**Objective.** To determine that the applicant:

1. Exhibits knowledge of the elements related to normal and crosswind takeoff and climb, including factors affecting performance, to include height/velocity information.
2. Establishes a stationary position on the surface or a stabilized hover, prior to takeoff in headwind and crosswind conditions.
3. Maintains RPM within normal limits.
4. Accelerates to manufacturer's recommended climb airspeed, ±5 knots.
5. Maintains proper ground track with crosswind correction, if necessary.
6. Remains aware of the possibility of wind shear and/or wake turbulence.
7. Completes the prescribed checklist, if applicable.

## B. TASK: NORMAL AND CROSSWIND APPROACH

REFERENCES: AC 61-13; Helicopter Flight Manual.

**NOTE:** If a calm wind weather condition exists, the applicant's knowledge of the crosswind elements shall be evaluated through oral testing; otherwise a crosswind approach and landing shall be demonstrated.

**Objective.** To determine that the applicant:

1. Exhibits knowledge of the elements related to normal and crosswind approach.
2. Considers performance data, to include height/velocity information.
3. Considers the wind conditions, landing surface, and obstacles.
4. Selects a suitable termination point.
5. Establishes and maintains the recommended approach angle, and rate of closure.
6. Remains aware of the possibility of wind shear and/or wake turbulence.
7. Avoids situations that may result in settling-with-power.
8. Maintains proper ground track with crosswind correction, if necessary.
9. Arrives at the termination point, on the surface or at a stabilized hover, ±2 feet (.6 meters).
10. Completes the prescribed checklist, if applicable.

## C. TASK: MAXIMUM PERFORMANCE TAKEOFF AND CLIMB

REFERENCES: AC 61-13; Helicopter Flight Manual.

**Objective.** To determine that the applicant:

1. Exhibits knowledge of the elements related to maximum performance takeoff and climb.
2. Considers situations where this maneuver is recommended and factors related to takeoff and climb performance, to include height/velocity information.
3. Maintains RPM within normal limits.
4. Utilizes proper control technique to initiate takeoff and forward climb airspeed attitude.
5. Utilizes the maximum available takeoff power.
6. After clearing all obstacles, transitions to normal climb attitude, airspeed, ±5 knots, and power setting.
7. Remains aware of the possibility of wind shear and/or wake turbulence.
8. Maintains proper ground track with crosswind correction, if necessary.
9. Completes the prescribed checklist, if applicable.

## D. TASK: STEEP APPROACH

REFERENCES: AC 61-13; Helicopter Flight Manual.

**Objective.** To determine that the applicant:

1. Exhibits knowledge of the elements related to a steep approach.
2. Considers situations where this maneuver is recommended and factors related to a steep approach, to include height/velocity information.
3. Considers the wind conditions, landing surface, and obstacles.
4. Selects a suitable termination point.
5. Establishes and maintains the recommended approach angle, (15° maximum) and rate of closure.
6. Avoids situations that may result in settling-with-power.
7. Remains aware of the possibility of wind shear and/or wake turbulence.
8. Maintains proper ground track with crosswind correction, if necessary.
9. Arrives at the termination point, on the surface or at a stabilized hover, ±2 feet (.6 meters).
10. Completes the prescribed checklist, if applicable.

## E. TASK: ROLLING TAKEOFF

**NOTE:** This TASK applies only to helicopters equipped with wheel-type landing gear.

REFERENCES: AC 61-13; Helicopter Flight Manual.

**Objective.** To determine that the applicant:

1. Exhibits knowledge of the elements related to a rolling takeoff.
2. Considers situations where this maneuver is recommended and factors related to takeoff and climb performance, to include height/velocity information.
3. Maintains RPM within normal limits.
4. Utilizes proper preparatory technique prior to initiating takeoff.
5. Initiates forward accelerating movement on the surface.
6. Transitions to a normal climb airspeed, ±5 knots, and power setting.
7. Remains aware of the possibility of wind shear and/or wake turbulence.
8. Maintains proper ground track with crosswind correction, if necessary.
9. Completes the prescribed checklist, if applicable.

# F. TASK: SHALLOW APPROACH AND RUNNING/ROLL-ON LANDING

REFERENCES: AC 61-13; Helicopter Flight Manual.

**Objective.** To determine that the applicant:

1. Exhibits knowledge of the elements related to shallow approach and running/roll-on landing, including the purpose of the maneuver, factors affecting performance data, to include height/velocity information, and effect of landing surface texture.
2. Maintains RPM within normal limits.
3. Considers obstacles and other hazards.
4. Establishes and maintains the recommended approach angle, and proper rate of closure.
5. Remains aware of the possibility of wind shear and/or wake turbulence.
6. Maintains proper ground track with crosswind correction, if necessary.
7. Maintains a speed that will take advantage of effective translational lift during surface contact with landing gear parallel with the ground track.
8. Utilizes proper flight control technique after surface contact.
9. Completes the prescribed checklist, if applicable.

# G. TASK: GO-AROUND

REFERENCES: AC 61-13; Helicopter Flight Manual.

**Objective.** To determine that the applicant:

1. Exhibits knowledge of the elements related to a go-around and when it is necessary.
2. Makes a timely decision to discontinue the approach to landing.
3. Maintains RPM within normal limits.
4. Establishes proper control input to stop descent and initiate climb.
5. Retracts the landing gear, if applicable, after a positive rate of climb indication.
6. Maintains proper ground track with crosswind correction, if necessary.
7. Transitions to a normal climb airspeed, ±5 knots.
8. Completes the prescribed checklist, if applicable.

## VI. AREA OF OPERATION: PERFORMANCE MANEUVERS

### A. TASK: RAPID DECELERATION

REFERENCES: AC 61-13; Helicopter Flight Manual.

**Objective.** To determine that the applicant:

1. Exhibits knowledge of the elements related to rapid deceleration.
2. Maintains RPM within normal limits.
3. Properly coordinates all controls throughout the execution of the maneuver.
4. Maintains an altitude that will permit safe clearance between the tail boom and the surface.
5. Decelerates and terminates in a stationary hover at the recommended hovering altitude.
6. Maintains heading throughout the maneuver, ±5°.

### B. TASK: 180° AUTOROTATION

REFERENCES: AC 61-13; Helicopter Flight Manual.

**Objective.** To determine that the applicant:

1. Exhibits knowledge of the elements related to a 180° autorotation terminating with a power recovery to a hover.
2. Selects a suitable touchdown area.
3. Initiates the maneuver at the proper point.
4. Establishes proper aircraft trim and autorotation airspeed, ±5 knots.
5. Maintains rotor RPM within normal limits.
6. Compensates for windspeed and direction as necessary to avoid undershooting or overshooting the selected landing area.
7. Utilizes proper deceleration, collective pitch application to a hover.
8. Comes to a hover within 50 feet (20 meters) of a designated point.

## VII. AREA OF OPERATION: NAVIGATION

### A. TASK: PILOTAGE AND DEAD RECKONING

REFERENCES: AC 61-21, AC 61-23, AC 61-84.

**Objective.** To determine that the applicant:

1. Exhibits knowledge of the elements related to pilotage and dead reckoning.
2. Correctly flies to at least the first planned checkpoint to demonstrate accuracy in computations.
3. Identifies and follows landmarks by relating the surface features to chart symbols.
4. Navigates by means of precomputed headings, groundspeed, and elapsed time.
5. Verifies the helicopter's position within 1 nautical mile (1.85 Km) of flight planned route at all times.
6. Arrives at the en route checkpoints within 3 minutes of the ETA.
7. Corrects for, and records, the differences between preflight fuel, groundspeed, and heading calculations and those determined en route.
8. Maintains the appropriate altitude, ±100 feet (30 meters) and established heading, ±10°.

### B. TASK: RADIO NAVIGATION AND RADAR SERVICES

REFERENCES: AC 61-21, AC 61-23, AC 61-84; Navigation Equipment Operation Manuals.

**Objective.** To determine that the applicant:

1. Exhibits knowledge of the elements related to radio navigation and ATC radar services.
2. Selects and identifies the appropriate facilities or coordinates, as appropriate.
3. Locates the helicopter's position relative to the navigation facilities or coordinates, as appropriate.
4. Intercepts and tracks a given radial or bearing.
5. Locates position using cross radials, coordinates, or bearings.
6. Recognizes and describes the indication of station or way point passage.
7. Recognizes signal loss and takes appropriate action.
8. Uses proper communication procedures when utilizing ATC radar services.
9. Maintains the appropriate altitude, ±100 feet (30 meters).

## C. TASK: DIVERSION

REFERENCES: AC 61-21, AC 61-23, AC 61-84.

**Objective.** To determine that the applicant:

1. Exhibits knowledge of the elements related to procedures for diversion.
2. Selects an appropriate alternate airport or heliport and route.
3. Promptly, diverts toward the alternate airport or heliport.
4. Makes an accurate estimate of heading, groundspeed, arrival time, and fuel consumption to the alternate airport or heliport.
5. Maintains the appropriate altitude, ±100 feet (30 meters) and established heading, ±10°.

## D. TASK: LOST PROCEDURES

REFERENCES: AC 61-21, AC 61-23, AC 61-84; AIM.

**Objective.** To determine that the applicant:

1. Exhibits knowledge of the elements related to lost procedures.
2. Selects the best course of action when given a lost situation.
3. Maintains the original or appropriate heading, and if necessary, climbs.
4. Attempts to identify nearest prominent landmark(s).
5. Uses available navigation aids and/or contacts an appropriate facility for assistance.
6. Plans a precautionary landing if deteriorating weather and/or fuel exhaustion is impending.

*FAA-S-8081-16* **Commercial**

## VIII. AREA OF OPERATION: EMERGENCY OPERATIONS

**NOTE:** Tasks F through I are knowledge only TASKS.

### A. TASK: POWER FAILURE AT A HOVER

REFERENCES: AC 61-13; Helicopter Flight Manual.

**Objective.** To determine that the applicant:

1. Exhibits knowledge of the elements related to power failure at a hover.
2. Determines that the terrain below the aircraft is suitable for a safe touchdown.
3. Performs autorotation from a stationary or forward hover into the wind at recommended altitude, and RPM, while maintaining established heading, ±5°.
4. Touches down with minimum sideward movement, and no rearward movement.
5. Exhibits orientation, division of attention, and proper planning.

### B. TASK: POWER FAILURE AT ALTITUDE

REFERENCES: AC 61-13; Helicopter Flight Manual.

**NOTE:** Simulated power failure at altitude shall be given over areas where actual touchdowns can safely be completed in the event of an actual powerplant failure.

**Objective.** To determine that the applicant:

1. Exhibits knowledge of the elements related to power failure at altitude.
2. Establishes an autorotation and selects a suitable landing area.
3. Establishes proper aircraft trim and autorotation airspeed, ±5 knots.
4. Maintains rotor RPM within normal limits.
5. Compensates for windspeed and direction as necessary to avoid undershooting or overshooting the selected landing area.
6. Terminates approach with a power recovery at a safe altitude when directed by the examiner.

## C. TASK: SYSTEMS AND EQUIPMENT MALFUNCTIONS

REFERENCES: AC 61-13; Helicopter Flight Manual.

**Objective.** To determine that the applicant:

1. Exhibits knowledge of the elements related to causes, indications, and pilot actions for various systems and equipment malfunctions.
2. Analyzes the situation and takes action, appropriate to the helicopter used for the practical test, in at least four of the following areas—

   a. engine/oil and fuel.
   b. hydraulic, if applicable.
   c. electrical.
   d. carburetor or induction icing.
   e. smoke and/or fire.
   f. flight control/trim.
   g. pitot static/vacuum and associated flight instruments, if applicable.
   h. rotor and/or antitorque.
   i. various frequency vibrations and the possible components that may be affected.
   j. any other emergency unique to the helicopter flown.

## D. TASK: SETTLING-WITH-POWER

REFERENCES: AC 61-13; Helicopter Flight Manual.

**Objective.** To determine that the applicant:

1. Exhibits knowledge of the elements related to settling-with-power.
2. Selects an altitude that will allow recovery to be completed no less than 1,000 feet (300 meters) AGL or, if applicable, the manufacturer's recommended altitude, whichever is higher.
3. Promptly recognizes and announces the onset of settling-with-power.
4. Utilizes the appropriate recovery procedure.

**E. TASK: LOW ROTOR RPM RECOVERY**

REFERENCES: AC 61-13; Appropriate Manufacturer's Safety Notices; Helicopter Flight Manual.

**NOTE:** The examiner may test the applicant orally on this TASK if helicopter used for the practical test has a governor that cannot be disabled.

**Objective.** To determine that the applicant:

1. Exhibits knowledge of the elements related to low rotor RPM recovery, including the combination of conditions that are likely to lead to this situation.
2. Detects the development of low rotor RPM and initiates prompt corrective action.
3. Utilizes the appropriate recovery procedure.

**F. TASK: DYNAMIC ROLLOVER**

REFERENCES: AC 61-13, AC 90-87; Helicopter Flight Manual.

**Objective.** To determine that the applicant:

1. Exhibits knowledge of the elements related to the aerodynamics of dynamic rollover.
2. Understands the interaction between the antitorque thrust, crosswind, slope, CG, cyclic and collective pitch control in contributing to dynamic rollover.
3. Explains preventive flight technique during takeoffs, landings, and slope operations.

**G. TASK: GROUND RESONANCE**

REFERENCES: AC 61-13; Helicopter Flight Manual.

**Objective.** To determine that the applicant:

1. Exhibits knowledge of the elements related to a fully articulated rotor system and the aerodynamics of ground resonance.
2. Understands the conditions that contribute to ground resonance.
3. Explains preventive flight technique during takeoffs and landings.

## H. TASK: LOW G CONDITIONS

REFERENCE: Helicopter Flight Manual.

**Objective.** To determine that the applicant:

1. Exhibits knowledge of the elements related to low G conditions.
2. Understands and recognizes the situations that contribute to low G conditions.
3. Explains proper recovery procedures.

## I. TASK: EMERGENCY EQUIPMENT AND SURVIVAL GEAR

REFERENCE: Helicopter Flight Manual.

**Objective.** To determine that the applicant:

1. Exhibits knowledge of the elements related to emergency equipment appropriate to the helicopter used for the practical test by describing—

   a. purpose of such equipment.
   b. location in the helicopter.
   c. method of operation.
   d. servicing requirements.
   e. method of safe storage.

2. Exhibits knowledge of the elements related to survival gear by describing—

   a. survival gear appropriate for operation in various climatological and topographical environments.
   b. location in the helicopter.
   c. method of operation.
   d. servicing requirements.
   e. method of safe storage.

*FAA-S-8081-16* **Commercial**

# IX. AREA OF OPERATION: SPECIAL OPERATIONS

## A. TASK: CONFINED AREA OPERATION

REFERENCES: AC 61-13; Helicopter Flight Manual.

**Objective.** To determine that the applicant:

1. Exhibits knowledge of the elements related to confined area operations.
2. Accomplishes a proper high and low reconnaissance.
3. Selects a suitable approach path, termination point, and departure path.
4. Tracks the selected approach path at an acceptable approach angle and rate of closure to the termination point.
5. Maintains RPM within normal limits.
6. Avoids situations that can result in settling-with-power.
7. Terminates at a hover or on the surface, as conditions allow.
8. Accomplishes a proper ground reconnaissance.
9. Selects a suitable takeoff point, considers factors affecting takeoff and climb performance under various conditions.

## B. TASK: PINNACLE/PLATFORM OPERATIONS

REFERENCES: AC 61-13; Helicopter Flight Manual.

**Objective.** To determine that the applicant:

1. Exhibits knowledge of the elements related to pinnacle/platform operations.
2. Accomplishes a proper high and low reconnaissance.
3. Selects a suitable approach path, termination point, and departure path.
4. Tracks the selected approach path at an acceptable approach angle and rate of closure to the termination point.
5. Maintains RPM within normal limits.
6. Terminates at a hover or on the surface, as conditions allow.
7. Accomplishes a proper ground reconnaissance.
8. Selects a suitable takeoff point, considers factors affecting takeoff and climb performance under various conditions.

# X. AREA OF OPERATION: POST-FLIGHT PROCEDURES

TASK: AFTER LANDING AND SECURING

REFERENCES: AC 61-13; Helicopter Flight Manual.

Objective. To determine that the applicant:

1. Exhibits knowledge of the elements related to after-landing procedures, including local and ATC operations, ramp safety, parking hand signals, shutdown, securing, and post-flight inspection.
2. Minimizes the hazardous effects of rotor downwash during hovering.
3. Selects a suitable parking area while considering wind and the safety of nearby persons and property.
4. Follows the recommended procedure for engine shutdown, securing the cockpit, securing rotor blades, and discharging passengers.
5. Performs a satisfactory post-flight inspection.
6. Completes the prescribed checklist, if applicable.